Environmental systems and societies

SECOND EDITION

TestPrep Workbook: DP Exam Practice

EXTEND education

Revise IB

A note from us

While every effort has been made to provide accurate advice on the assessments for this subject, the only authoritative and definitive source of guidance and information is published in the official subject guide, teacher support materials, specimen papers and associated content published by the IB. Please refer to these documents in the first instance for advice and guidance on your assessments.

Any exam-style questions in this book have been written to help you practise and revise your knowledge and understanding of the content before your exam. Remember that the actual exam questions may not look like this.

Dr Emma M. Shaw

Rose Githinji

Second Edition Editor: Nigel Gardner

SL
Standard Level

Published by Extend Education Ltd, Alma House, 73 Rodney Road, Cheltenham, UK, GL50 1HT

www.extendeducation.com

The right of Dr Emma M. Shaw and Rose Githinji to be identified as authors of this work has been asserted by them with the Copyright, Designs and Patents Act 1988.

Second Edition Editor and Contributor: Nigel Gardner

Typesetting by York Publishing Solutions Pvt. Ltd., INDIA

Cover photo by Sander Wehkamp, Unsplash.com

First published 2019

Second Edition published 2022

26 25 24 23 22

10 9 8 7 6 5 4 3 2 1

ISBN 978-1-913121-38-9

Author acknowledgements

Dr Emma M. Shaw: For my dad, Phil Shaw, you will forever be my inspiration.

Rose Githinji: I'd like to thank my students who inspire me. Special thanks to two people that always support and encourage me: Sam and Brian.

Nigel Gardner: I would like to thank every ecologist before me. Every ESS teacher that has lent me ideas. Every student that has road tested many many resources. La garrigue juste pour être là. But especially my parents and family who've let me be the ecologist I am.

Copyright notice

Other considerations

A reminder that Extend Education is not affiliated with the International Baccalaureate.

Many people have worked to create this book. We go through rigorous editorial processes, including separate answers checks and expert reviews of all content. However, we all make mistakes. So if you notice an error in the paper, please let us know at info@extendeducation.co.uk so we can make sure it is corrected at the earliest possible opportunity.

If you are an educator with a passion for creating content and would like to write for us, please contact info@extendeducation.co.uk or write to us through the contact form on our website www.extendeducation.co.uk.

CONTENTS

HOW TO USE THIS BOOK

This excellent exam practice book has been designed to help you prepare for your Environmental Systems and Societies Standard Level (SL) exams. It is divided into three sections.

EXPLAIN

The EXPLAIN section gives you a rundown of your paper, including the number of marks available, how much time you'll have and the assessment objectives (AOs) and command terms. In this section, you will be able to self-assess against the command terms to find out which questions you might need more practice in.

SHOW

The SHOW section gives you some examples of different questions you will come across in the exam. It is designed to help you learn the question types and the kinds of answers you can give to get you the maximum number of marks.

TEST

This is your chance to try out what you've learned. The TEST section has full sets of exam-style practice papers filled with the same type and number of questions that you can expect to see in the exam.

Set A
Paper 1 and Paper 2

Presented with a lot of tips and guidance to help you get to the correct answer and boost your confidence.

Set B
Paper 1 and Paper 2

Presented with fewer helpful suggestions and hints so you need to make sure you have done some revision before attempting these papers.

Set C
Paper 1 and Paper 2

Presented with even fewer helpful suggestions and hints so you need to make sure you have revised enough before attempting these papers.

Set D
Paper 1 and Paper 2

Presented with space for your notes and no hints – the perfect way to test if you are exam ready.

Use these papers early on in your revision...   **Use these papers closer to the exam...**

There are **ANSWERS** at the back of the book so you can check how you did in your practice papers.

Features

Take a look at some of the helpful features in these books that are designed to support you as you do your practice papers.

These will point you in the direction of the right answer!

These are general hints for answering the questions.

These are referred to as AOs all the way through this book

This box reminds you of the assessment objective being tested.

Beware of making common and easy-to-avoid mistakes!

These flag up common or easy-to-make mistakes that might cost you marks.

The command terms are like a clue to how you should answer your questions

COMMAND TERMS

These boxes outline what the command term is asking you to do.

These boxes contain really useful advice about what examiners are looking for

ANSWER ANALYSIS

These boxes include advice on how to get the most possible marks for your answer.

KNOWING YOUR PAPER

Part of taking any timed exam is understanding the format and questions. This will shorten the time you need to sort out what is expected from you in the exam. This section will help to make sure there are no surprises! The table below outlines the structure, how much time to spend on each section and how many questions to answer.

There are instructions in this section for self-assessing using an assessment objectives table. This is to help you find out where you are already in your revision.

How are you assessed?

You will sit **two** written papers.

Paper 1* 🖩	Paper 2* 🖩
Case study: Interpretation and evaluation of data presented on a specific, previously unseen case study	**Section A:** Short-answer questions **Section B:** You choose two of four essay-style questions in your Paper 2 exams (however, when you do the practice papers in this book, you will have to choose two questions from three options instead of four)
Weighting 25%	Weighting 50%
35 marks	65 marks
1 hour	2 hours

*Students who speak English as an additional language can have an approved translation dictionary.

Note: Your individual investigation is weighted at 25%.

Your assessment objectives

There are **four** assessment objectives for Environmental Systems and Societies (ESS). Assessment objectives 3 and 4 share command terms, though assessment objective 4 uses them within the Internal Assessment. The table below shows the assessment objectives that are relevant to your exams – AO1, AO2 and AO3. Make sure you are clear on what you need to demonstrate for each one.

Assessment objective 🎯	Command terms	Which questions test this?	Examples
Assessment objective 1	Define Draw Label List Measure State	Questions in the exam that test your understanding of AO1 are written to check your knowledge and understanding of facts, concepts, and methodologies.	**State** the index used to indirectly measure pollution levels in freshwater environments. **[1 mark]**
Assessment objective 2	Annotate Apply Calculate Describe Distinguish Estimate Identify Interpret Outline	Questions in the exam that test your understanding of AO2 will ask you to use information to answer a question. These questions may ask you to use data that is provided, or knowledge of different concepts, to support a theory or argument.	Using named examples, **distinguish** between a primary and secondary air pollutant. **[2 marks]**

Make sure that you know exactly what to expect on the day of the exam.

Remember to write in blue or black ink only. You will also need a ruler, pencil, eraser and access to a calculator.

Read all questions carefully. Underline key elements to make sure you answer what is asked.

Remember you can use diagrams to help illustrate your answers.

Write your answers on the lines provided.

Make sure you can read what you write clearly; if the examiner cannot read it, then they cannot mark it.

Assessment objective 🎯	Command terms	Which questions test this?	Examples
Assessment objective 3	Analyse Compare and contrast Construct Deduce Discuss Evaluate Explain Examine Justify Suggest To what extent	Questions in the exam that test your understanding of AO3 will ask you to explain the reasons for something. You may be asked for case studies or examples that support your answer.	**Evaluate** the effectiveness of a named human population control strategy in decreasing birth rates. **[5 marks]**

AO3 and AO4 share command terms but only some are likely to be used in your exams (the AO3 terms shown in the table). The missing objectives you are not likely to find in your exams are: Comment, Demonstrate, Derive, Design, Determine, Predict, Sketch.

How do you know what you know?

The three assessment objectives 1–3 are levelled – with increasingly complex answers required for the questions in each.

- **Assessment objective 1** demonstrates recall-type knowledge.
- **Assessment objective 2** demonstrates descriptive knowledge and application.
- **Assessment objective 3** demonstrates the use of evaluation and synthesis (joining more than one idea together).

You can think about these as a ladder. The higher up the ladder you go, the wider and more detailed your understanding of the concepts and issues in ESS becomes.

Assessment objective ladder

Question focus	Approximate grade level	Quality of understanding
Synthesis	7	Assessment objective 3
Evaluation	6	
Application	5	Assessment objective 2
Description	4	
Definition	3	Assessment objective 1
Statement	2	
	1	

In both Paper 1 and Paper 2, just over 50% of the paper assesses assessment objective 3. Therefore, to do well in the papers, you need to be prepared for AO3 questions.

How can you know how well prepared you are?

Learning the ESS topic content is a key part of exam revision. However, it is not the only part. There is another way of assessing how prepared you are for your exams. You can use the assessment ladder to help you understand which areas you need to work on and which areas you already have a good foundation in.

Whenever you answer a set of questions for ESS, they will use the **command terms**. By looking at which command terms you have answered correctly, you can formatively self-assess where you are on the assessment ladder.

Using the self-assessment table below will help.

Self-assessment AO table

Assessment objective	Command term	What level am I?	
3	Analyse, Compare and contrast, Construct, Deduce, Discuss, Evaluate, Explain, Examine, Justify, Suggest, To what extent	I correctly answered **MORE** than 50% of questions that test this assessment objective.	
		I correctly answered **LESS** than 50% of questions that test this assessment objective.	
2	Annotate, Apply, Calculate, Describe, Distinguish, Estimate, Identify, Interpret, Outline	I correctly answered **MORE** than 50% of questions that test this assessment objective.	
		I correctly answered **LESS** than 50% of questions that test this assessment objective.	
1	Define, Draw, Label, List, Measure, State	I correctly answered **MORE** than 50% of questions that test this assessment objective.	
		I correctly answered **LESS** than 50% of questions that test this assessment objective.	

(Source: Adapted from an idea by Stephan Taylor, i-biology.net)

We are now going to look at two examples to help you understand how to use this table.

Example: Student 1

This student has just completed a practice Paper 1 as part of their revision. They have marked their paper and filled out the self-assessment table.

Assessment objective	Command term	What level am I at?	
3	Analyse, Compare and contrast, Construct, Deduce, Discuss, Evaluate, Explain, Examine, Justify, Suggest, To what extent	I correctly answered **MORE** than 50% of questions that test this assessment objective.	
		I correctly answered **LESS** than 50% of questions that test this assessment objective.	✓
2	Annotate, Apply, Calculate, Describe, Distinguish, Estimate, Identify, Interpret, Outline	I correctly answered **MORE** than 50% of questions that test this assessment objective.	✓
		I correctly answered **LESS** than 50% of questions that test this assessment objective.	
1	Define, Draw, Label, List, Measure, State	I correctly answered **MORE** than 50% of questions that test this assessment objective.	✓
		I correctly answered **LESS** than 50% of questions that test this assessment objective.	

In this example, the student has ticked both the **more than 50%** boxes for assessment objective 1 and 2 but the **less than 50%** box for assessment objective 3. This suggests that the student has very good understanding of the topics that the questions covered up to assessment objective 1 and 2, but is developing the use of assessment objective 3 to answer questions. Using the ladder, you can then approximate what level this might be.

> Let's approximate the level for this particular student... Using the ladder, if student 1 took their IB ESS exams tomorrow, they would be on track for a Grade 5 and starting to get into the Grade 6 area. However, with a bit more practice in mastering AO3 questions, they will solidly be into Grade 6 and maybe even start to reach a Grade 7.

Example: Student 2

This student has just completed a practice Paper 2 as part of their revision. They have marked their paper and filled out the self-assessment table.

Assessment objective	Command term	What level am I at?	
3	Analyse, Compare and contrast, Construct, Deduce, Discuss, Evaluate, Explain, Examine, Justify, Suggest, To what extent.	I correctly answered **MORE** than 50% of questions that test this assessment objective.	
		I correctly answered **LESS** than 50% of questions that test this assessment objective.	✓

Assessment objective	Command term	What level am I at?	
2	Annotate, Apply, Calculate, Describe, Distinguish, Estimate, Identify, Interpret, Outline	I correctly answered **MORE** than 50% of questions that test this assessment objective.	
		I correctly answered **LESS** than 50% of questions that test this assessment objective.	✓
1	Define, Draw, Label, List, Measure, State	I correctly answered **MORE** than 50% of questions that test this assessment objective.	✓
		I correctly answered **LESS** than 50% of questions that test this assessment objective.	

Using the ladder, if student 2 took their IB ESS exams tomorrow, they would be on track for a Grade 4 with some evidence that they are developing higher order skills, but in a limited way. With a bit more practice in mastering both AO2 and AO3 questions, they could achieve a Grade 5 or even a Grade 6.

This student has ticked **more than 50%** on assessment objective 1 but then **less than 50%** on both assessment objectives 2 and 3. This suggests that the student is probably answering questions that get well into the assessment objective 2 band, but still needs to work on these AOs as well as AO3. Again, the ladder can be used to approximate where this student's level is right now.

This is a useful way of not only helping you know which topics you may be stronger or weaker in, but also focusing on what **type of questions** you might need to work on more.

Checking what you know

Before using the full papers in this book, you can check your confidence level with questions that test different assessment objectives. To do this, arrange any questions you have received in class by topic, and then use the self-assessment table and the assessment ladder to help you understand where you are right now, in ESS.

Your past questions are also a great way of revising the entire course.

Self-assessment challenge

Next are 28 questions arranged in topic order. Try them out and then use the self-assessment table at the end to see where your performance currently sits.

Your answers for these self-assessment questions are on pages 21–24.

This task is just about testing how well you can answer different question types (questions using different command terms). It is not about marks.

ANSWER ANALYSIS

These questions do not have marks assigned to them and there is no time limit. You will need to assess your answers against the markscheme on page 21 and will just need to decide if your answers are **partially correct** or **fully correct**. You can use extra paper to answer the questions if needed. You will then put a tick in the marking table if you decide that your answer is fully correct.

Remember your performance is only on this set of questions. You need to do a lot of questions to make sure you have covered the whole course. That is where the question paper sets in this book will help.

Topic 1

1. **Suggest** how historical events and figures have been influential in shaping the current environmental movement.

....

....

....

....

....

....

....

....

....

....

....

....

....

....

....

....

....

Topic 1 and 2

2. **Describe** a historical event that led to the current understanding of biomagnification and bioaccumulation.

....

....

....

....

....

....

....

....

....

....

....

Topic 2

Figure 1: Survivorship curves for different species

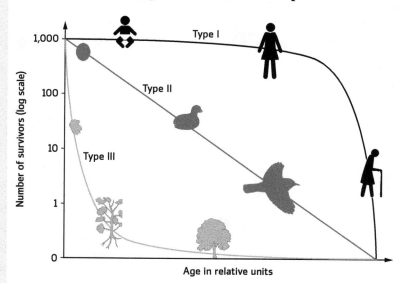

3. Species found along a succession can be classified according to how quickly they reproduce. Figure 1 (above) shows the types of survivorship curves for different species.

State the type of species represented by the following curves:

Type I

....

Type III

....

@ AO1

Topic 2

4. **Explain** the use and procedure for each of the following:
   - Sweep nets
   - Pitfall trap
   - Pooter

@ AO3

....

....

....

....

....

....

....

....

....

....

....

....

Topic 1 and 2

5. **Draw** a systems diagram to summarize the pathways of energy through an ecosystem.

Topic 2

6. **Explain** how the transfers of energy and matter through photosynthesis support the first and second laws of thermodynamics.

....

....

....

....

....

....

....

....

....

....

....

....

....

....

....

....

....

Topic 2

7. **Discuss** the human impacts on the carbon and nitrogen cycles.

⊙ **AO3**

....
....
....
....
....
....
....
....
....
....
....
....
....
....
....
....
....
....

Topic 3

8. **Discuss** the benefits and limitations of the IUCN Red list's role in protecting global biodiversity.

⊙ **AO3**

....
....
....
....
....
....
....
....
....
....
....
....
....
....

Topic 3

9. **Define** each of the following forms of biodiversity:

    (a) Habitat diversity

    ....

    ....

    (b) Genetic diversity

    ....

    ....

⊙ **AO1**

Topic 3

10. Extinction of species can be either by natural causes or human causes. **Explain** one natural cause.

    ....

    ....

    ....

    ....

⊙ **AO3**

Topic 4 and 5

11. **Suggest** two reasons why food is in short supply in some societies.

    ....

    ....

    ....

    ....

⊙ **AO3**

Topic 1 and 5

12. With respect to one of the major Environmental Value Systems, **discuss** a potential solution to decrease the environmental footprint of animal husbandry.

    ....

    ....

    ....

    ....

    ....

    ....

    ....

    ....

    ....

    ....

⊙ **AO3**

Topic 5

13. Soil formation involves both transfer and transformation processes.

◉ **AO2**

   (a) **Identify** two of these transfer processes.

....

....

   (b) **Identify** two of these transformation processes.

....

....

Topic 5

Figure 2: A profile of soil horizons

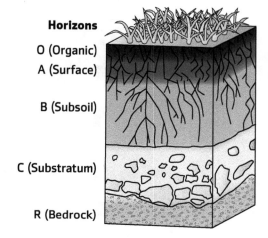

Horizons
O (Organic)
A (Surface)
B (Subsoil)
C (Substratum)
R (Bedrock)

14.

   (a) **State** the soil horizon that contains the most nutrients.

◉ **AO1**

....

   (b) **State** two factors that determine the primary productivity of a soil.

....

....

Topic 5

15. Soil degradation is an increasing global problem.

Outline two main reasons for this.

◉ **AO2**

....

....

Topic 1 and 6

16. Many of the issues related to atmospheric pollution are as a result of the burning of fossil fuels in industry, transportation, and daily life.

Evaluate possible solutions that would either reduce the need for, or reduce the release of air pollution from, fossil fuel combustion. Address each of the three named areas where combustion often occurs and give your opinion as to which you feel will be the most effective measure and why.

⌖ AO3

Topic 6

17. Distinguish between stratospheric and tropospheric ozone.

⌖ AO2

Topic 6

18. **Explain** the causes of stratospheric ozone depletion and the role of the Montreal Protocol in reducing these effects.

◎ **AO3**

....

....

....

....

....

....

....

....

....

....

Topic 6

19. **Identify** two sources of each of the greenhouse gases shown:

Carbon dioxide

◎ **AO2**

....

....

Nitrous oxide

....

....

Topic 7

Figure 2: Changes in the mass of ice sheets found in Antarctica

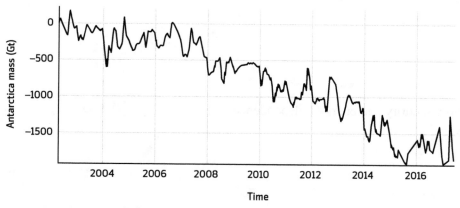

(Source: https://climate.nasa.gov/vital-signs/ice-sheets/)

20. Describe the data shown in Figure 2.

Topic 1 and 7

21. Describe, using examples, how feedback loops are part of the ongoing issues related to the reduction in ice mass on the planet.

Topic 1 and 7

22. Outline the success of a current energy production or saving project that is being implemented somewhere around the world. Determine which Environmental Value System (EVS) lens is being used to develop this method.

Topic 7

23. Outline the link between energy security and energy independence.

Topic 7

24. Using examples to support your answer, **explain** factors that can influence the energy choices of a country.

◎ AO3

....

....

....

....

....

....

....

....

....

....

....

....

....

Topic 8

25. Define the natural increase in a human population providing the relevant equation.

◎ AO1

....

....

....

....

Topic 1 and 8

26. Plastic is proving to be one of the most persistent anthropogenic pollutants in our systems. **Discuss** potential waste management strategies to help address this problem.

◎ AO3

....

....

....

....

....

....

....

....

....

Topic 1 and 8

27. Explain the dynamic nature of the concept of natural capital, using a specific example.

....

....

....

....

....

....

....

....

....

....

....

....

....

Topic 8

28. There are different types of Solid Domestic Waste (SDW), and different strategies for managing it.

With refence to landfill and incineration, **evaluate** the efficiency of these two approaches to SDW management.

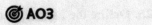

....

....

....

....

....

....

....

....

....

....

....

....

....

Answers

Here are the answers to the 28 self-assessment questions. When you are comparing your answers with the markscheme, you will need to make a judgement about whether your answer is partially correct, or fully correct. Tick every answer you think you got fully correct in the table below. Count up how many questions for each assessment objective you got fully correct and put the totals at the bottom of the table. Then fill out the assessment objective table on page 24.

Go back to the assessment ladder. Where would you put your current performance? Which question types do you need to practice more?

Question	AO1	AO2	AO3
1			
2			
3			
4			
5			
6			
7			
8			
9			
10			
11			
12			
13			
14			
15			
16			
17			
18			
19			
20			
21			
22			
23			
24			
25			
26			
27			
28			
Fill out your number of correct answers →	/5	/9	/14

> Remember, it **does not matter** if you get less than 50% of the questions right for any of the assessment objectives when doing the exam papers in this book. It is just important to know where you are so that you can practise the right **types of questions** that can help you to get a better grade.

> You might wonder why we are putting so much emphasis on AO3 questions as part of your self-assessment. Because 50% of the assessment in Paper 1 and Paper 2 is from AO3 questions, focusing on those makes a big difference. All AO3 questions contain AO2 and AO1 within them. For example: Explain (AO3) is 'give a detailed account with reasons'. This includes Describe (AO2) which is 'give a detailed account', but describe also contains State (AO1) 'give a brief answer'. The difference when answering the question completely as an Explain question is that you have to STATE the problem, DESCRIBE what is happening and EXPLAIN the reasons why it is happening.

Answers

1. There are several major historical events and figures that are still influential within the environmental community today.

   Rachel Carson wrote the book *Silent Spring*, which highlighted the impact of pesticides on our environments. As a result, more attention has been paid to the way that pollutants can move and magnify through the food chain. Prior to this, the release of methyl mercury from the Crisco factory in Minamata, Japan, presented the first major record of this biomagnification and bioaccumulation. This knowledge has influenced the development of pesticides and other toxic substances.

   Action taken by Greenpeace and other similar NGOs has shown that through the power of people it is possible to influence change, such as through the 'Save the Whales' campaign in the 1990s.

   In conclusion, most of the data we have to develop models and theories and drive decision making is from past disasters and environmental issues. For example, previous flooding events have allowed the development of early warning systems to help minimize the impact of similar disasters.

2. Minamata or DDT and Silent Spring

   Minamata: Methyl mercury was released from the Chisso factory into Minamata bay. Cats showed the symptoms first, then humans, which came from eating the fish and subsequent mercury bioaccumulation.

   DDT: This pesticide was developed to help eradicate malaria. In the countryside, songbirds were reduced (this is why the book is called *Silent Spring*). Shells of eggs in raptors (birds of prey) were thinner so breeding was unsuccessful as biomagnification enhanced the impacts. Therefore, the predators were worst affected.

3. Type I: *K*-strategists
   Type III: *r*-strategists

4. All three are methods of sampling small animals. They are methods used to capture small motile organisms to estimate their abundance using Lincoln's Index.

   Sweep net – a funnel-shaped net with a long handle; it is swept back and forth through vegetation; insects captured in the net are counted.

   Pitfall trap – a hole is dug into the ground; a suitable container, e.g. a plastic cup, is placed in the hole so that its rim is at the same level as the soil surface; crawling insects and other animals fall into the pitfall trap and are counted.

   Pooter – a small jar with a lid and two tubes penetrating through the lid/stopper; one tube has a filter attached at the end; the second tube goes into a collecting chamber; sucking on the first tube draws the insect into the collecting chamber.

5.

6. • 1st law of thermodynamics: Energy is not created nor destroyed but is changed from one form to another; plants trap light energy from the sun and convert it to chemical energy through photosynthesis.
   • 2nd law of thermodynamics: Entropy of a system increases over time; of the solar energy, only a small percentage is used in photosynthesis; some is reflected from leaves; some light doesn't get into chloroplasts; some is of different wavelengths; inefficiency of photosynthesis.

7. • Biogeochemical cycles involve an equilibrium that balances elements in the environment.
   • Phosphorus and nitrate fertilizers – much are lost through run-off; both result in eutrophication, leading to algal bloom which prevents sunlight reaching other plants; death of fish and other aquatic animals from lack of oxygen; decomposition increases, which releases carbon dioxide.
   • Sewage from sewage treatment plants increases organic matter in water bodies; faeces are fed on by decomposers releasing ammonium compounds which are further converted to nitrates by nitrifying bacteria, resulting in more eutrophication.
   • Combustion of fossil fuels releases carbon dioxide – a greenhouse gas that contributes to global warming and climate change. Clearing of vegetation, which acts as a carbon sink, results in the increase in carbon dioxide in the atmosphere.
   • Livestock ranching releases ammonia from their waste; nitrogen enters the soil system and the hydrological system through ground water, run-off and leaching.
   [Your answer would need to discuss at least 3 of the 4 areas to reach the highest grade band.]

8. The International Union for the Conservation of Nature's Red list is the most comprehensive project assessing the conservation status of a large proportion of global species.

   Pros:
   • Identifies and protects the most vulnerable species.
   • Funds in-situ and ex-situ research to help improve the status of endangered species.
   • Has the legal right to stop the trade and movement of certain plants and animals.
   • Raises awareness, is an international agreement.

   Cons:
   • There are only a small number of species in some of the groups that have been assessed.
   • Requires a lot of data collection to keep it up to date.
   • The amount of information depends on the amount of funding and researchers.
   • Needs to be supported by local initiatives and consistent law enforcement.

9. (a) Habitat diversity – range of different habitats per unit area in an ecosystem or biome.
   (b) Genetic diversity – total number of genetic characteristics of a species within a population.

10. Natural causes: Volcanic eruptions; Earthquakes; Landslides; Drought; Tsunami. These are out of anyone's control even though humans could indirectly cause some of them. For example, drought as a result of deforestation.

11. Any two from:
    • Conflict/wars/minefields destroy agriculture infrastructure
    • Differences in soil/climate/water availability
    • Lack of preservation/storage facilities in LEDCs
    • Lack of distribution infrastructure in some countries
    • Lack of political incentives to increase food production
    • Natural hazards, e.g., tsunamis, hurricanes, volcanic eruptions, droughts
    • Human impact, e.g., overgrazing/overfishing, increased desertification due to climate change

12. Any one of the following:

    Ecocentric
    This is the ecological view that is holistic in respect to the rights of nature and its ability to repair itself.
    • **Idea:** Develop intercropping systems that incorporate growing grain and vegetables along with the development of productive field margins to offset some of the footprint of the meat industry.
    • **Pros:** Increased production of a variety of crops from the same area and potential natural pest control, as a result of a greater plant diversity in the area.
    • **Cons:** Makes harvesting of grain and other crops more time consuming and could result in further depletion of soil fertility.

    Technocentric
    This is a view that to develop, the environment needs technological innovation, and that resource replacement overcomes resource depletion.
    • **Idea:** Growing artificial meat in laboratories.
    • **Pros:** It reduces the land needed for grazing and growing feed and creates a reduction in methane production from cattle.
    • **Cons:** It is likely to increase the cost of meat and uncertainty regarding the safety of artificially produced products.

    Anthropocentric
    This view looks at people as the environmental managers of global sustainability. Human population control can be key in this view.
    • **Idea:** Implement additional tax on meat and provide subsidies to farmers who increase their production of vegetable crops.
    • **Pros:** Reduce meat consumption and therefore reduce the amount of land needed to raise livestock and the emissions related to them.
    • **Cons:** It could make meat an elite product that is only accessible by the wealthy and could be viewed as a loss of human rights to choose what to eat.

13. **Transfers**
    • Movement of water
    • Deposition of sediments by erosion
    • Leaching – minerals dissolved in water move through soil
    • Translocation – movement of soil particles in suspension, biological mixing

    Transformations
    • Decomposition weathering
    • Nutrient cycling
    [Your answer needs to include two answers for each process.]

14. (a) Horizon O
    (b) Mineral content, water holding capacity, drainage, potential to retain organic matter, air spaces.

15. Any two from:
    • Deforestation
    • Intensive agriculture
    • Urbanization
    • Agricultural practices such as irrigation and monoculture

16. **Industry**
    The need for fossil fuels in industry can be reduced by introducing solar or wind power to provide some of the energy requirements, or by increasing the efficiency of the process.

    The release of air pollution from industry can be reduced by adding scrubbers to chimney stacks to remove some of the pollutants, and by improving the technology for cleaning.

Transportation

The need for fossil fuels can be reduced by improving mass public transport systems, improving clean technologies, and encouraging walking or cycling by pedestrianizing city centres.

The release of air pollution from transportation can be reduced by increasing engine efficiency, developing cleaner fuels, and using renewable power such as electric cars.

Daily life

To reduce the need for fossil fuels in your daily life take public transport, walk or cycle, share cars to work, work from home, improve natural lighting in your home and improve home insulation.

Reduce the release of air pollution in your daily life by servicing vehicles regularly, only driving when necessary, combining activities to reduce transport needs, reducing use of heating and lights, etc.

Opinion

In my opinion, I feel that dealing with the transportation sector is most important as it touches on parts of industry and daily life. This is an area where lots of developments are already taking place, e.g., electric cars. But further improvements will encourage better transportation systems, greener vehicles and more sustainable behaviour.

17.

	Stratospheric ozone	Tropospheric ozone
Location	Also known as the ozone layer that surrounds the planet	Lower parts of the troposphere, often starting at ground level
Cause / formation	Ultraviolet (UV) rays interact with oxygen to create ozone	Primary pollutants such as NO_x and VO_x react with sunlight to create ozone
Effect on environment	Protects environment from harmful UV (B and C) rays damaging plants and animals	Produces polluted air, often in cities, that can result in smog, reduced visibility and brown-coloured air
Effect on humans	Protects humans from harmful UV rays that can cause skin cancer or cataracts	Causes respiratory problems, sore eyes and skin; and can cause death in the young, elderly or sick

18. Stratospheric ozone depletion is caused by ozone depleting gases (ODGs) such as chlorofluorocarbons (CFCs) and other halogenated gases. CFCs come from refrigerator units, air conditioners and previously all aerosol spray cans. These very slowly make their way up into the stratosphere where ultraviolet (UV) light causes a chlorine atom to break away from the halogenated gas. This then binds to one of the oxygens in ozone (O3) reducing the thickness of the ozone layer.

The Montreal Protocol in 1987 called for a phasing out of production of non-essential ODGs and replacement with non-harmful alternatives. It has been the most effective environmental agreement so far, and there has been evidence of repairs on the damaged ozone layer and a successful reduction in the production of GHGs.

19. Carbon dioxide – any two from:
- Burning fossil fuels
- Respiration by living organisms
- Breakdown of organic material
- Volcanic vents
- Forest fires/bushfires

Nitrous oxide – any two from:
- Fuel combustion
- Wastewater management
- Industrial processes

20. Rapidly increasing reduction in ice mass with over 1500Gt lost between 2004 and 2016.

21. Feedback is one of the processes that either maintains a balance in a system or amplifies change.

Positive feedback can be related to the melting of ice. As it melts, the albedo in that area will change and more heat will be absorbed than reflected. This will cause more heat to be absorbed by the Earth's surface, therefore further increasing the amount of ice that will melt. This cycle moves away from the original equilibrium and creates a constantly changing level of equilibrium.

22. A technocentric lens is appropriate here. Waste to energy systems have been developed around many of the Scandinavian countries for some time. A lot of energy is generated through the incineration of waste in order to generate power. As this does not further deplete fossil fuel supplies, this is an effective alternative for producing large amounts of energy. However, there are a number of major environmental issues relating to this generation of energy as it still produces a significant amount of air pollutants when incineration takes place. This is a very technocentric way of dealing with this issue as it is completely reliant on the incineration of waste, despite the fact that it is currently not as efficient as it could be.

23. Energy independence comes from making strategic choices that are influenced by the availability of suitable energy production sources. These will differ in each country due to differences in location, topography, development and resource availability. Mountainous areas might prefer to look at wind power or hydroelectric dams, while coastal regions might want to employ tidal power or wave power. Without energy independence there is no way to achieve energy security.

24. Energy choices can be influenced by a number of factors including cultural, economic, technologically driven, and related to availability.

Availability of a natural way to produce energy is very important as this helps to create a sustainable source. For a country with a lot of moving water – for example, Bhutan – choosing hydroelectricity is the most strategic response. Many coastal countries, such as the UK, make use of the prominent coastal winds to help develop energy independence.

Technology is a key factor that is also related to the level of development in a country. MEDCs such as the USA have the financial flexibility to invest in technology development, whereas many of the African nations do not have that flexibility and are therefore limited in the choices that they can make.

25. Natural increase (%) = (birth rate per 1000 – death rate per 1000)/10 Natural increase is the percentage that the human population changes, per year, in a given area.

26. Pollution should be tackled on three levels, reducing production and release, and clean-up when a pollution event takes place. Plastics are in so many parts of our lives that new, durable replacements need to be developed and used that have similar properties as plastic but are able to be broken down in the environment. For example, using cloth bags for groceries and other shopping, using glass jars and bottles instead of plastic, using reusable straws, having a refillable water bottle. However, these all still produce waste of some variety even though it is less toxic than plastic.

To reduce the release of plastics into the environment there needs to be better recycling facilities and schemes. Some countries are cutting the release of waste plastics by developing plastic payment schemes that allow people to use plastic bottles as cash for bus rides, movie tickets etc. This is taking place in Beijing, Surabaya Indonesia, and Sydney Australia. Other countries are using stuffed plastic bottles as bricks for ecobuildings and they have even been made to make small canoes. These schemes are excellent but need government support and financial backing for use to become widespread.

Despite all the reduction schemes, there is already a lot of plastic in the environment that needs to be removed. The Great Pacific Garbage patch is a huge area in the ocean where plastics and other garbage accumulate. A device called the Seabin has been developed that sucks water in and filters out the plastics. Despite its benefits there are possibilities of mortalities as surface animals may be sucked into the machine.

27.
- Value of a resource is dynamic because its status changes over time.
- Resource only a resource if it is useful to people.
- Technological development means resources used change. Resources become more valuable as new technologies need them. For example, flint was once used as a hand tool and was an important resource then. It is no longer used as it was replaced by metal extraction from ores. Therefore, it is not a resource now.
- Uranium had little value before development of nuclear energy. Now, it is an important resource in nuclear energy production.
- Resources currently in use may cease to be resources in future as development of technology continues.

28. Landfill

Strength:
- Cheap to set up

Weaknesses:
- Land will eventually run out.
- Increased health problems for those living near landfills. For example, heart problems, birth defects.
- Pollution of air by methane from decomposing waste – methane is a GHG (greenhouse gas).
- Pollution of water and soil by heavy metals and chemicals.
- Area used up could have been turned into a nature reserve or lake.
- Communities near the site of the landfill are usually opposed to it being set up.

Incineration

Strengths:
- Heat from burning can be used to generate electricity.
- Reduces volume of waste by up to 90%.
- Safe for disposing hazardous waste such as clinical waste.

Weaknesses:
- Pollution from carbon dioxide, sulphur dioxide, nitrogen dioxide, nitrous oxide, and particulates lead to acid rain, smog, and lung disease.
- Volume of traffic getting waste to the incinerator increases, further increasing air pollution.
- Increased accidents from increased traffic.
- Ash resulting is toxic and still needs disposal in a landfill.
- Initial cost of building incinerators is high.

Now you have counted up your correct answers, fill out this blank self-assessment table. Where would you put your current performance on the ladder on page 6?

Assessment objective	Command term	What level am I at?	
3	Analyse, Compare and contrast, Construct, Deduce, Discuss, Evaluate, Explain, Examine, Justify, Suggest, To what extent.	I correctly answered **MORE** than 50% of questions that test this assessment objective.	
		I correctly answered **LESS** than 50% of questions that test this assessment objective.	
2	Annotate, Apply, Calculate, Describe, Distinguish, Estimate, Identify, Interpret, Outline	I correctly answered **MORE** than 50% of questions that test this assessment objective.	
		I correctly answered **LESS** than 50% of questions that test this assessment objective.	
1	Define, Draw, Label, List, Measure, State	I correctly answered **MORE** than 50% of questions that test this assessment objective.	
		I correctly answered **LESS** than 50% of questions that test this assessment objective.	

To find out more information about how you can self-assess, scan the QR code to get your marking tables to use with this book.

Before you begin your practice exams

The ESS syllabus covers a wide range of concepts and you will often be asked to relate several of them together. To make sure that you are able to confidently answer the questions, you need to understand the command terms, the specific ESS vocabulary and a range of case studies you can use to support ideas. You also need a clear understanding of the following topics.

Put a tick in each box when you are happy that you have fully studied that topic, when you are confident with the definitions, and then when you feel that you have enough case studies and examples to support that topic.

Topic checklist

Topic	Studied	Definitions	Case studies
1. Foundations of environmental systems and societies	☐	☐	☐
2. Ecosystems and ecology	☐	☐	☐
3. Biodiversity and conservation	☐	☐	☐
4. Water and aquatic food production systems and societies	☐	☐	☐
5. Soil systems, terrestrial food production systems and societies	☐	☐	☐
6. Atmospheric systems and societies	☐	☐	☐
7. Climate change and energy production	☐	☐	☐
8. Human systems and resource use	☐	☐	☐

Make a table of case studies, aspects of topics and where they can be used. This will help you to work out if you have enough examples or if you need to learn more.

Some historical case studies relate to more than one area of the syllabus. For example, Rachel Carson's *Silent Spring* (1962) shows:
- pollution impacts
- bioaccumulation and biomagnification
- food chain effects
- population ecology
- an ecocentric method to clean up polluted areas.

Know what the command terms are asking you to do. Look out for the tips in this book.

Remember that this is a holistic subject. This means topics are all interlinked in some way, even though they are presented separately.

What to do in your exam

Check details of the exam are correct

Write in your candidate session number on the front of your paper

Focus for a moment, take a deep breath and open the paper

Familiarize yourself with the command terms in the questions to make sure you know what the question wants you to do.

Read the instructions carefully before starting

For **Paper 2** make a quick plan in your head for your two chosen questions. When the reading time is over, quickly write down your ideas next to the question

Look through the paper to get an idea of what to expect

Before attempting a question, read the question carefully. What does the command term want you to do?

Look at the number of marks allocated for the question, as this will help you know how much you have to write

Try to give yourself enough time at the end of the exam to check over your answers

If you get stuck on a question, move on but make sure that you come back to it

Attempt every question (except for Part B of Paper 2)

If you finish with time to spare, go over the paper again to make sure you did not miss anything

 Make sure that you keep an eye on time.

 Paper 1 is a series of sets of information that are all related to one case study. You must use data from the Resource Booklet AND your own knowledge to answer the questions in Paper 1.

ANSWER ANALYSIS

For **Paper 2**, make sure you use an example or case study if asked for one: without it you will not get the full marks.

ANSWER ANALYSIS

When reading data from a graph, use a ruler and pencil to mark and measure to get as accurate an answer as you can.

Check that the results of your calculations make sense, and that you use units of measurement when needed.

SHOWING WHAT YOU KNOW

In this section, some model student answers have been shown to give you an idea of the type of answer you could give in the exam. Within these there will be tips, traps and advice that will help you work out why the answers are correct. Command terms will be defined throughout. Please note that in many cases in ESS there is more than one possible answer. Make sure you support your answer when it is clear there is more than one potential answer.

Paper 1 examples

Welcome to Paper 1. The questions for this paper will be based on an unseen case study that you will find in a Resource Booklet. You will have one hour to complete this paper and can get a total of 35 marks. Take a look at some example answers to the types of questions you get in Paper 1.

ANSWER ANALYSIS

The Resource Booklet may not give you direct answers to questions but it will help you to remember and apply some concepts.

While most of the answers are in the Resource Booklet, some questions may require application of general knowledge and common sense.

Remember this is a new case study, not any of the ones you have studied. Apply knowledge and understanding gained to this case study.

One case study does not test knowledge on one topic. It cuts across many topics in the course.

Figure 1: Geographical map of Sweden

(Credit: iStock, PeterHermesFurian)

Figure 2: Population pyramid of Sweden from 2016

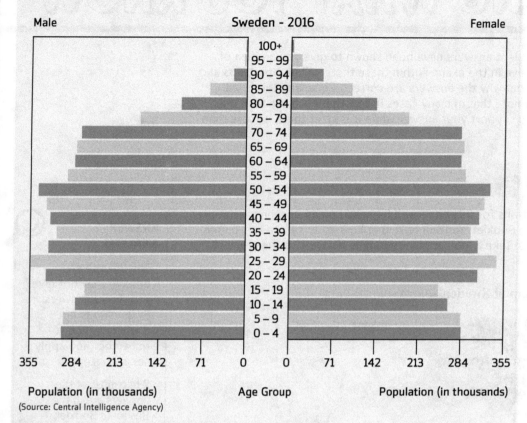

(Source: Central Intelligence Agency)

Figure 3: Fact file for Swedish population and energy

<u>**Human population**</u>

10,040,995 people in 2018

Dependency ratio: 58.5

Youth dependency: 27.4

Elderly dependency: 31.1

Birth rate: 12.1/1000 population

Death rate: 9.4/1000 population

Urban population in 2018: 87.4%

<u>**Electricity generation**</u>

Fossil fuels: 5%

Nuclear: 22%

Hydropower: 41%

Other renewables: 32%

(Source: Central Intelligence Agency)

1. Using Figures 1 and 3, **state** the geographical distribution of the Swedish population. [2]

Most of the major cities are found on the coast and 87% of the Swedish population are found in urban habitats. Therefore, most of Sweden's population live on the coast.

STATE
Give a brief answer without explanation or calculation.

ANSWER ANALYSIS

This student would receive full marks. They have used information from both of the named figures to create a detailed response.

2. Using Figures 2 and 3, **explain** the current population dynamics with particular reference to the impacts of dependency. [4]

Life expectancy in Sweden is very high with both males and females living to their 90s. High populations are found up to the age of around 70 for both males and females. There is a decrease in population growth evidence for a number of years, as demonstrated by the narrowing base of the pyramid.

Dependency is where individuals are not of the age to be able to work and earn money, and therefore look after themselves. Sweden has a very high level of dependence with both young and elderly individuals being reliant on 41.5% of the population to support them.

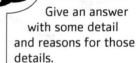

EXPLAIN
Give an answer with some detail and reasons for those details.

ANSWER ANALYSIS

This student has answered exactly what the question wants. They first describe what is happening with the population then use the idea of dependency to help justify the answer that has been given.

3. Energy production is important for any country in the world. Using Figure 3, **evaluate** the provision for power generation in Sweden. [5]

Sweden is a country that has quite good provision for its own energy production, meaning there is quite high energy independence. With only 5% of production coming from fossil fuels, Sweden has a good development of a sustainable renewable system through hydropower (41%) and other renewables (32%). The rest (22%) comes from nuclear power, which is specifically well adapted for the coastal location. Although good, there is still a heavy dependency on hydropower that could result in power generation issues should there be some problem with the water in Sweden, e.g. if the area became much hotter and droughts began to happen.

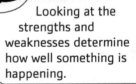

EVALUATE
Looking at the strengths and weaknesses determine how well something is happening.

ANSWER ANALYSIS

This student has given a good answer. They have provided an overall statement about the level of power production, with reasons why the choices that have been made are the right ones.

Paper 2 (Section A) examples

Section A

1. **Figure 1: Changes in sea level (mm) from 1880 to present day**

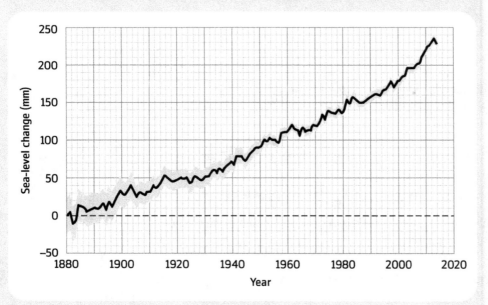

(Source: https://climate.nasa.gov/climate_resources/133/sea-level-historic-data/)

(a) Outline the pattern of sea-level change both before and after 1986. **[1]**

Before:

From 1880–1986 the sea level is steadily increasing but with regular fluctuations from one year to the next. It rose from 0 mm to around 150 mm in 106 years.

After:

From 1986–2013 the increase was faster with very few downward fluctuations. It rose from 150 mm to around 235 mm in 27 years.

(b) **Construct** a systems diagram to describe one of the possible feedback mechanisms that may have attributed to the patterns observed in Figure 1. [4]

Feedback – positive feedback

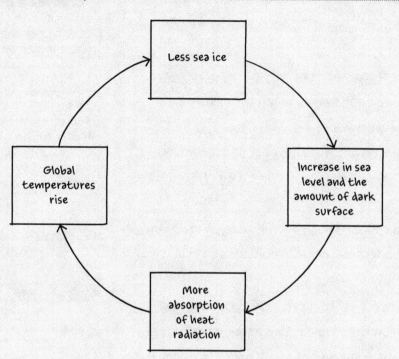

ANSWER ANALYSIS

For this course you must know how to put the environmental concepts in the form of a systems diagram.

A systems diagram has storages in the form of boxes and flows in the form of arrows. when considering transfers and transformations. For feedback systems, the diagram should have step-by-step effects that feed into the next effect.

(c) The use of fossil fuels is often presented as one of the major causes of climate change. **Evaluate** one renewable and one non-renewable energy source in relation to the impact it potentially has on the climate. [4]

	Coal	Solar energy
Implications	Finite supply so it will run out in the near future	Pollution-free energy source
Limitations	Produces harmful greenhouse gases	Technology still needs developing to reduce the space required to mass-generate energy

Avoid using nuclear power in this type of question. It is not considered to be a renewable energy as the uranium is depleted. It is considered to be an **alternative energy**.

This student has used a table to lay out their answer. This ensures all areas of the question are covered and makes it easier for the examiner to pick out the information.

(d) **Discuss** how effective international climate change agreements have been in reducing the anthropogenic impacts on climate change. [7]

1 Kyoto Protocol 1997 to cut greenhouse gas emissions:

The agreement focused on industry, but it was hard to get countries to commit to effecting major decreases in greenhouse gases.

Global emissions have increased since the agreement was signed, but it is a well-known event that helped many understand problems associated with greenhouse gas emissions.

2 The Paris Agreement 2015 to decarbonize energy production:

The Paris Agreement was effective as it was signed by 174 countries and the European Union.

It allowed flexibility as no country had specific goals to reach by certain dates, allowing the less-developed countries to still take part but develop at the same time.

There were no penalties imposed if countries failed to take any measures, which allowed some to exploit this agreement.

Despite the lack of perceived success of the Kyoto Protocol, it opened the door for future agreements and highlighted the importance of regulation to reduce emission levels.

2. **Define** species. [1]

A group of organisms that can interbreed and create fertile offspring.

ANSWER ANALYSIS

This student has addressed both the **positive** and **negative** sides of each of the two agreements they have presented.

This question addresses AO3 because it asks for reasons and a conclusion supported by evidence.

There are other examples that could be used in this situation.

If you do not include the word **fertile** this statement is no longer correct, as it is possible for two different species to breed and produce non-fertile offspring. For example, a donkey and a horse produce a mule, which is not fertile and can therefore not reproduce.

3. Using specific named examples, **explain** two upland soil conservation management techniques giving specific reasons for their use. [4]

- Upland soil is often needed to grow crops and so managing the soil is very important. It is key that the soil keeps its nutrients as well as keeping a good depth.

- Contour ploughing is used to grow and irrigate crops using the natural movement of the water along the contours of the hill. This stops surface soil being removed in heavy rains.

- Terracing is a way of creating flat areas down a steep slope so that crops such as rice can be grown in upland habitats. This allows water and nutrients to be kept in the area.

This student has included an answer that is very similar to the one given by the International Baccalaureate (IB). Learning the key points of **definitions** in the IB will help you provide the correct answers.

ANSWER ANALYSIS

This student would be awarded the maximum 4 marks here as two techniques are named, briefly described and their benefits clearly explained.

You do not have to give a named example of where this technique is being used, just the technique itself.

Paper 2 (Section B) examples

The following is an example of one of the three-part essay style questions that make up Section B of Paper 2. In this section you will be given four sets of questions from which you need to choose two. Each group of questions follows the same pattern. The first part is worth 4 marks, the second part 7 marks and the final part is worth 9 marks.

1. The conservation status of an individual species is assessed in relation to many factors.

   (a) **Identify** two factors of concern in relation to protected species. **[4]**

   > If a species has a small population size, then this will make it more vulnerable to genetic inbreeding. If there are only a few populations of that species, they are at risk of being wiped out by a natural disaster or disease.

   (b) Habitat management is often more effective than species management.

   Explain how managing a habitat can be of greater biodiversity value than managing areas for a single species. Use named examples to support your answer. **[7]**

   > Managing an area for one species only manages the conditions that are optimal for that species with less attention being paid to the rest of the organisms. For instance, species-based conservation for the Giant Panda would have some of the focus on the conservation of bamboo to ensure that it was accessible and abundant enough to support and sustain a developing population. This may be a disadvantage to species that are not reliant on bamboo.
   >
   > Habitat-based conservation works on a more holistic basis, preserving the integrity of the entire habitat. The Galapagos islands are an archipelago that is protected as a whole site. Despite there being a number of highly protected species, the conservation does not concentrate on any one of them. One of the main conservation methods is the control and limiting of visitor numbers to the island. The conservation in this area includes the culture, heritage and traditions of the islands.
   >
   > Therefore, management on a habitat level will result in a greater level of biodiversity than managing for an individual species.

Part (a) is worth 4 marks and addresses AO2.

Part (b) is worth 7 marks and addresses AO3.

Part (c) is worth 9 marks and addresses AO3.

This question is only worth 4 marks. Do not be fooled by the label of 'essay style questions'; you do not have to write long paragraphs to get the full marks.

ANSWER ANALYSIS

The answer for part (a) would be awarded full marks as each factor is identified in relation to its impact on that species.

ANSWER ANALYSIS

For part (a), only naming the factors will give 2 marks.

EXPLAIN

To describe with examples.

ANSWER ANALYSIS

This student would have received full marks for this answer as they have addressed both management areas in relation to their impact on biodiversity.

(c) Species-based conservation is often carried out in two complimentary ways to ensure that the species will not become extinct.

With reference to a named species, **evaluate** in-situ and ex-situ conservation strategies discussing which you feel is the most effective method of protecting a species. **[9]**

<u>Golden lion tamarin monkey:</u>

This is a critically endangered species that has had an improved status through the use of both in-situ and ex-situ management techniques. In-situ conservation is related to intervention within the habitat of the species. Ex-situ conservation looks at conducting research and conservation outside of the animal's habitat. This often takes place in zoos, nature reserves, etc.

I feel that to be truly effective both strategies need to be employed. If a species is removed from its habitat in order to facilitate breeding in a captive setting with the idea that individuals may, in the future, be reintroduced to the wild, it is clear that the conservation efforts have not addressed the causes of the initial decline. If these causes are not dealt with then the species will struggle to thrive when reintroduced, as before.

Habitat conservation is essential to ensure that the species has a stable base from which to develop. Without some of the key habitat requirements for certain species, it is difficult to manage an area for one species. However, if an area was conserved in terms of vegetation, for example, then that will help develop a solid foundation for the monkey's population to base itself on. Within that habitat, the species has a function and it is of paramount importance that the interactions that take place are taken into account. For instance, if the correct canopy structure and balance of available food items are present, it is clear that these factors will also benefit other animal species that use or rely on that food source. Equally, golden lion tamarin monkeys are involved with seed dispersal in their habitat, and so have an impact on the lower and the higher trophic levels in that ecosystem.

ANSWER ANALYSIS

This student has given very clear definitions between the two conservation types with good identification of the differences.

ANSWER ANALYSIS

This answer scored highly because mentions the importance of both techniques, but it did not get full marks as the student did not identify which of these they thought was most effective.

This question addresses AO3, because an evaluation is needed.

TESTING WHAT YOU KNOW

Set A

In this section you will be able to test yourself with different sets of exam practice papers under exam conditions. By taking these practice papers, you will build your confidence and identify any areas you need a bit more practice on. The papers that are coming up in Set A have a lot of additional guidance in the margin to help you get to the right answer. So attempt this set first.

All you need is a book, a timer, a pen and some extra paper to use if you run out of answer lines. Then you can check your answers in the back of the book when you are done.

Take a deep breath, set your timer, and good luck!

Paper 1

Instructions for this paper state that you must use data from the Resource Booklet **AND** your own knowledge to answer the questions.

- Answer **all** questions. Answers must be written on the answer lines provided.
- Set your timer for 1 hour
- There are 35 marks available
- The Resource Booklet provides information on Northern Thailand. Use the Resource Booklet and your own studies to answer the following.

1. Using Figure 1(c) **list** <u>two</u> biomes found in Thailand. **[1]**

2. Using Figure 1(b) and Figure 2, **state** the topography surrounding Chiang Mai. **[1]**

3. Figure 3(a) shows information about the Northern Thailand hill tribes. **Identify** <u>two</u> factors that may have led to a loss of biodiversity in the region. **[2]**

Your Resource Booklet for this paper is on page 47.

LIST

Just really simply list what's been asked for. You don't need to put any other description or example.

ANSWER ANALYSIS

You can choose any two biomes from the Resource Booklet but you **must** use the exact wording from the booklet. There is only one mark here so if you don't use the exact wording for both, you will get zero marks.

STATE

Briefly name the topography. Don't give any other explanation.

IDENTIFY

Provide answers from a selection of possible sources.

For Q3, be careful you don't repeat the same answer just with different wording. You need to include different factors.

4. Using Figure 3(b), **calculate** the rate of population change between 2010 and 2020. [1] **[1]**

5. Figure 4 shows information about biodiversity around Chiang Mai. **Identify two** reasons why the forests around Chiang Mai are considered important for biodiversity. **[2]**

6. **Identify one** of the criteria that may have been used by the International Union for Conservation of Nature (IUCN) Red List to classify Asian Elephant as endangered. **[1]**

7. Using Figure 5, **justify** a choice of **one** species as the most suitable to promote conservation. **[3]**

8. Using Figures 6(a) and (b), **explain** how fragmentation and reduced male home-range in their mating season may negatively impact on an improvement in the current IUCN status of the Great Hornbill. **[4]**

💬 **EXPLAIN**

To link the two factors to potential reasons why there is no improvement in their status.

9. Using Figures 7(a) and 7(b):

(a) **Outline two** benefits of burning rice straw. **[2]**

....

....

....

....

(b) **Outline two** disadvantages of burning rice straw. **[2]**

....

....

....

....

10. Using Figures 7 and 8, **describe** the process of thermal inversion. **[2]**

....

....

....

....

11. Using Figure 9, **describe** the annual pattern of PM_{10} readings between 2000 and 2012. **[1]**

....

....

....

....

12. Using all the information from the case study in the Resource Booklet:

(a) **Suggest two** reasons why air pollution has become worse over recent years. **[2]**

....

....

....

....

(b) **Explain** how the problem of pollution from the burning of rice stubble can be addressed. **[2]**

....

....

....

....

13. Outline <u>one</u> advantage of increased ecotourism on species conservation. **[1]**

14. Outline <u>two</u> disadvantages of increased ecotourism on species conservation. **[2]**

Make sure that your answer is targeted specifically at the impact on the conservation of the species, not on the general disadvantages of ecotourism.

15. To what extent could development away from traditional rice farming for the Hill peoples around Chiang Mai lead to greater sustainability? **[6]**

Set A: Paper 2

- Set your timer for **2 hours**
- Section A – answer **all** the questions
- Section B – answer **two** questions
- The maximum mark for this examination paper is **65 marks**.

Question 1 (a) tests AO1 as it uses the word 'state'.

Section A

1. **Figure 1: Primary Succession Following Glacial Retreat**

(Source: N Gardner – Four Corners Education)

(a) **State** whether succession is a temporal or spatial pattern of change. **[1]**

STATE
To give a brief name, value or statement without any explanation.

Look at the key words **spatial** (space) and **temporal** (time).

(b) **Outline two** reasons why climax communities are more stable than pioneer or intermediate communities. **[2]**

There are more than two possible answers here. If you write more than two, the examiner will only mark the first two. So make sure you give the answers you are most secure with.

You need to give **two** reasons.

(c) **Describe** how primary productivity changes as communities move from pioneer to climax. **[2]**

Remember that primary productivity is related to plant growth in a given area. Look at the diagram for guidance.

(d) **Outline two** reasons why the species in the pioneer community are more likely to be *K*-strategists rather than *r*-strategists. **[2]**

OUTLINE
Your two reasons here should be brief as this is an outline question. Just one sentence for each reason will be enough.

(e) Zonation is an example of spatial distribution, **state** an abiotic factor that leads to zonation. [1]

2. **Figure 2: National percentage energy consumption from different fuels in 2011**

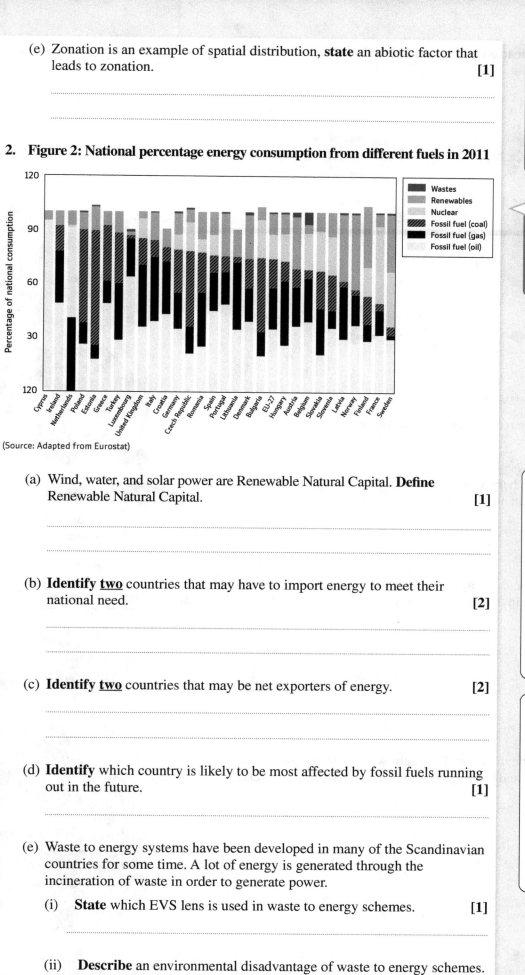

(Source: Adapted from Eurostat)

(a) Wind, water, and solar power are Renewable Natural Capital. **Define** Renewable Natural Capital. [1]

(b) **Identify <u>two</u>** countries that may have to import energy to meet their national need. [2]

(c) **Identify <u>two</u>** countries that may be net exporters of energy. [2]

(d) **Identify** which country is likely to be most affected by fossil fuels running out in the future. [1]

(e) Waste to energy systems have been developed in many of the Scandinavian countries for some time. A lot of energy is generated through the incineration of waste in order to generate power.

(i) **State** which EVS lens is used in waste to energy schemes. [1]

(ii) **Describe** an environmental disadvantage of waste to energy schemes. [1]

3. **Figure 3: Biomagnification and bioaccumulation of DDT in an arctic biome community**

Bioaccumulation

Year 1   Year 2

Biomagnification

(Source: N Gardner – Four Corners Education)

(a) **State** the main source of energy in this biome. [1]

....

(b) Using Figure 3 to help, **distinguish** between biomagnification and bioaccumulation in an ecosystem. [1]

....

....

(c) DDT is an insecticide sprayed on food crops as an aerosol. **Describe** how it could be found in Arctic ecosystems. [2]

....

....

....

....

....

(d) Using the example in Figure 3, **explain** why organisms most at risk from bioaccumulation and biomagnification are in higher tropic levels. [5]

....

....

....

....

....

....

....

....

....

Question 3(a) is asking you to use knowledge about the wider concept of the biome and not directly about biomagnification or bioaccumulation.

You may not have studied this as an example in class, so you need to be able to transfer what you have learned to a new situation.

DISTINGUISH
Provide clear differences between two or more things, here these are biomagnification and bioaccumulation.

Think about what happens if DDT particles get into the upper atmosphere.

If you speak English as an additional language and you do not know the definition of these words, you may not be able to find them in a dictionary. Instead break the word down. Bio means it is related to something living, then look up the end of the word.

STATE the problem, **DESCRIBE** what is happening and **EXPLAIN** the reasons.

Section B

Choose any **two** from the list of three questions. (Remember, in the real exam you will choose two from a list of four.)

4. (a) **Distinguish** between *r* and *K* selected species. [4]

....

....

....

....

....

....

....

You need to mention both number of offspring and use of resources to get the full marks.

Remember that this part of the topic is looking at survivorship and which factors impact on that.

AO2 is being assessed, as the command term 'distinguish' is used.

(b) As a habitat moves through successional stages, **explain** how the numbers of *r* and *K* species will change in relation to these stages. [7]

....

....

....

....

....

....

....

....

....

....

....

....

....

....

ANSWER ANALYSIS

One mark is given for the definitions, with three marks for each strategy (including details of the trait the species has and how it makes it better adapted). A maximum of four marks will be given if only one strategy is addressed.

AO3 is being assessed, as an explanation is needed.

Think about which stage of succession each strategist would be best suited to. Go back to your answers for part (a) if you are unsure.

(c) **Discuss** how ecosystem stability, succession and human activity are connected. [9]

....

....

....

....

....

ANSWER ANALYSIS

For this question, you may need to make connections between different parts of the ESS course. The 9-mark questions often expect you to synthesize what you know from different topics.

....
....
....
....
....
....
....
....
....
....

5. (a) Construct an **annotated** diagram of the soil profile. **[4]**

ANNOTATE
To provide labels and information on a diagram.

ANSWER ANALYSIS

Include labels and information. A plain diagram will only receive one out of four marks.

This question does not ask for a diagram of a particular soil type, just a general diagram. Also your diagram does not have to be a work of art, but does need to be very clear.

(b) Referring to your diagram in part (a), **explain** the process of leaching and how elements such as soil type, slope and vegetation can influence the severity of the leaching. **[7]**

Start your answer with a definition of leaching.

Do not just list as many soil conservation techniques as possible. Without detail you will not get all the marks available.

....
....
....
....
....
....

Answer lines continue on next page

....

....

....

....

....

....

....

....

(c) Human activities such as monocultures, overgrazing, overuse of chemical fertilizers and water extraction can degrade soils.

With reference to a commercial farming system you have studied, **evaluate** a soil management strategy used to conserve soils. **[9]**

....

....

....

....

....

....

....

....

....

....

....

....

....

....

....

....

6. (a) **Outline** how development policies can influence population growth. **[4]**

....

....

....

....

....

....

....

....

....

(b) **Explain** changes in human population dynamics due to changes in birth rates and fertility rates. Use named examples to support your reasoning. [7]

....
....
....
....
....
....
....
....
....
....
....
....
....
....
....
....

(c) 'The human population has exceeded the planet's carrying capacity.'

Discuss how this statement applies to a local population you have studied. [9]

....
....
....
....
....
....
....
....
....
....
....
....
....
....
....
....

EXPLAIN

Offer a possible answer. There is not just one right answer.

If you are also a geography student, do not waste time by including information not required in ESS.

AO3 is being addressed in this question, as a suggestion with examples is asked for.

Present a balance of changes, some positive and some negative. Try to include a named country, rather than a group of countries such as less economically developed countries (LEDC) or more economically developed countries (MEDC).

Start by defining 'carrying capacity', and use examples.

Do not only present one side of the argument.

Think about the major human-made pressures on the planet that might impact on the carrying capacity. This could include development rates, medical care, education and contraception. Think how these differ in developing and developed countries.

ANSWER ANALYSIS

Your answer needs to discuss how successfully your community has applied measures.

RESOURCE BOOKLET

This Resource Booklet is for use with Set A, Paper 1 on page 36.

Set A: Paper 1

Figure 1(a)
The location of Chiang Mai, Thailand

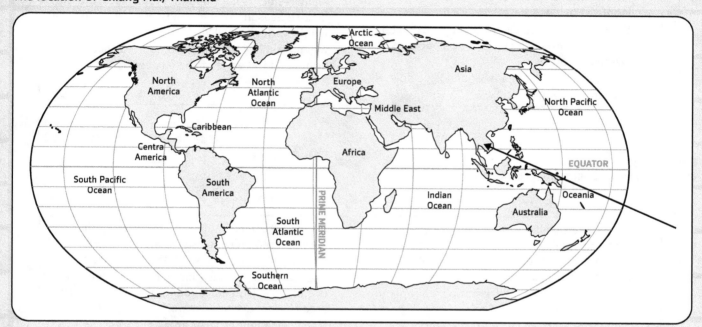

Figure 1(b)
Topographic map of Thailand, showing the location of Chiang Mai

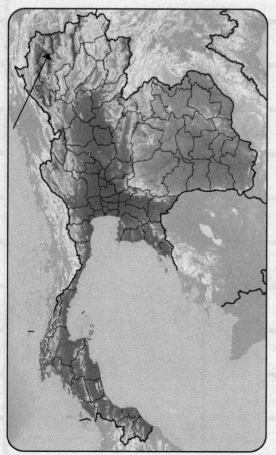

Figure 1(c)
Main vegetation types in Thailand

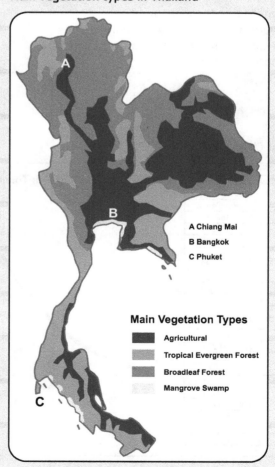

A Chiang Mai
B Bangkok
C Phuket

Main Vegetation Types

Agricultural
Tropical Evergreen Forest
Broadleaf Forest
Mangrove Swamp

(Source: Dr Brains, www.maps-for-free.com, GNU Free Documentation License)

Figure 2
General fact file on Chiang Mai

- Chiang Mai city is located in Northern Thailand and is the second largest city in Thailand.
- Chiang Mai city has approximately 201,000 residents and is a cultural hub for Northern Thailand.
- Car ownership rose significantly since 2013 due to a first time buyer's scheme encouraging people to buy new cars.
- It is 300 m above sea level, and approximately 69% of the province consists of mountains covered in forest.
- Chiang Mai city sits in a bowl and is completely surrounded by mountains.

- Chiang Mai has three official seasons:
  - rainy season from June to October
  - hot season from March to May
  - 'cold' season from December to early February, when temperatures can drop below 10 °C at night.
- Many residents have added a fourth season, the 'Smoke Season' from March to May (although the length varies annually).
- Chiang Mai has a rapidly growing international airport on the edge of the city centre.

Figure 3(a)
Northern Thailand hill tribe fact file

- There are a number of hill tribes located in Northern Thailand. The six main tribes (Karen, Hmong, Akha, Lahu, Lisu, Mien) are sub-divided into smaller tribes.
- Each tribe occupies a different part of upland Northern Thailand
- Income comes from shifting cultivation agriculture and tourism.
- Maize is a dominant commercial crop, but most communities produce the majority of their own food within the mountains.
- Tourism through treks and homestays is becoming an important source of income for these communities.

- Many of these communities are made up of unregistered individuals who are therefore limited in the work in which they can take part.
- The standard of education in these communities is quite low, with most young people leaving the area to attend school or work in the city.
- Many communities clear areas of forest for agriculture; due to increasing populations, the area needed to support them needs to increase.
- Many communities are part of opium replacement crops schemes, developing systems to grow cash crops like ornamental flowers, strawberries, coffee, cabbages and macadamia nuts.

Figure 3(b)
Population of Chiang Mai, 1920–2020

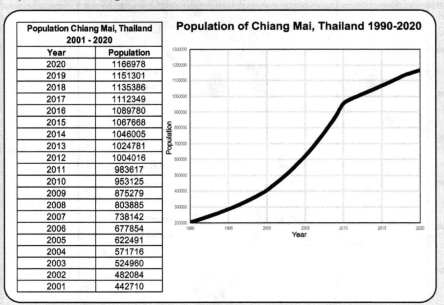

Population Chiang Mai, Thailand 2001 - 2020	
Year	Population
2020	1166978
2019	1151301
2018	1135386
2017	1112349
2016	1089780
2015	1067668
2014	1046005
2013	1024781
2012	1004016
2011	983617
2010	953125
2009	875279
2008	803885
2007	738142
2006	677854
2005	622491
2004	571716
2003	524960
2002	482084
2001	442710

Population of Chiang Mai, Thailand 1990-2020

(Source: Data extracted from World Population Review – Chiang Mai (worldpopulationreview.com/world-cities/chiang-mai-population))

Figure 4
Fact file on biodiversity in Chiang Mai province

- Chiang Mai province is in the IUCN Indo Burma hotspot.
- The major habitats are upland rainforest and agriculture in the form of rice, maize and fruit tree plantations.
- The biodiversity of the area is high, but reducing due to habitat fragmentation.
- Great Hornbill, classified by IUCN as 'Vulnerable'.
- *Anasedulio majophrae,* endemic but abundant species of frog.
- Green-tailed sunbird, endemic to the area.
- *Bhutanitis lidderdalli ocellatomaculata,* a butterfly species endemic to Chiang Mai.

- *Astraeus hygrometricus,* a hygroscopic earthstar mushroom, which is considered a delicacy and can be sold for a high price.
- *Dendrobium chiangdaoense,* an endemic epiphytic orchid.
- Indochinese tigers (*Panthera tigris tigris*), exact numbers are not available although the wild population
- numbers are known to be very low. This species is classified as 'Endangered' on the IUCN Red list.
- Asian elephants (*Elephas maximus*), there are a few wild individuals still remaining but exact numbers are not known. This species is classified as 'Endangered' on the IUCN Red list.

Figure 5
Some of the species found in the Chiang Mai province

(a) Asian elephant (*Elephas maximus*)

(Source: David Stanley, Elephants in Kaudulla Wewa, flickr.com (CC BY 2.0))

(b) Indochinese tiger (*Panthera tigris tigris*)

(Source: Frida Bredesen on Unsplash)

(c) Hygroscopic earthstar mushroom (*Astraeus hygrometricus*)

(Source: Björn S., Geastrum sp., flickr.com (CC BY-SA 2.0))

(d) Local Dendrobium epiphytic orchid (*Dendrobium chiangdaoense*)

(Source: Swallowtail Garden Seeds, Dendrobium junceum. An epiphytic orchid native to Borneo and the Philippines, flickr.com)

Figure 6(a)
Great Hornbill (Bucero bicornis) fact file

- The Great Hornbill is a mostly frugivorous and relatively large (95–130 cm length and up to 4 kg) bird found in Thailand.
- The Great Hornbill is classed as 'Vulnerable' on the IUCN Red list.
- Males have a home-range of around 14.7 km, which reduces to around 3.7 km in the mating season.

- They prefer non-logged, mature forest in hilly regions and are dependent on large areas of undisturbed forest.
- Roosting occurs on the tallest trees and nesting takes place in the mature forest.
- They are important dispersers of seeds.

Figure 6(b)
A Great Hornbill at a nest in a tree

(Source: Lip Kee, Great Hornbill (Buceros bicornis), flickr.com (CC BY-SA 2.0))

Figure 7(a)
Chiang Mai 'Smoke Season' fact file

Reasons

- Farmers burning rice and maize straw – despite a burning ban between January and April.
- Cold nights and hot daytime temperatures cause temperature inversion in the valley at night.
- Very still weather conditions with minimal air movement.
- *Astraeus hygrometricus* is a hygroscopic earthstar mushroom that grows underneath leaf litter in forests. Burning makes them easier to find.
- Other fungus species rely on burning to stimulate growth.
- Car ownership has grown exponentially in the last 5 years.
- There are impacts from burning taking place in neighbouring countries.

Impacts

- The 'Smoke Season' is an annual event that lasts between 2 weeks and 2 months between late February and mid-April.
- Air pollution ($PM_{2.5}$ levels) can reach a maximum of 183. The world's safe standard is 50.
- Visibility is severely impaired. In 2015 all planes were grounded at Chiang Mai airport for a number of days.
- Residents are advised to wear n95 filter masks when outside, reduce overall time outside and limit any aerobic exercise.
- Advice is to use air purifiers in homes with air conditioning and windows closed.
- Sporting events are stopped and schools keep students inside for the majority of the day.
- Tourism falls greatly and some businesses close during this time.
- Many locals and expats leave the area during this time.

Solutions

- Localised burning ban from January to April.
- Collecting, baling and selling stubble.
- Ploughing rice straw back in to the soil as an organic amendment.
- Rice straw collected and used for mulching and mushroom media.
- Collecting rice straw and creating Biochar, a carbon-rich charcoal used as a valuable soil amendment. Biochar is made without the production of harmful gases.

(Source: Solutions extracted from: Kanokkanjana K. and Garivait S. (2013) Alternative Rice Straw Management Practices to Reduce Field Open Burning in Thailand. International Journal of Environmental Science and Development, 4(2): 119–123.)

Figure 7(b)
Subsistence agriculture process showing the rotation of growing, harvesting and burning of rice and maize that often takes place twice a year

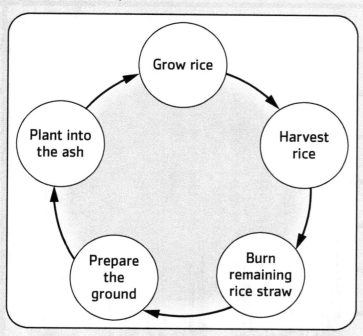

Figure 8
View of Doi Suthep mountain on the west side of Chiang Mai city. The top image (a) is during 'Smoke Season' and the lower image (b) is the same location during 'Rainy Season'

(Source: Dr Emma M. Shaw)

Figure 9
PM$_{10}$ levels (*particles < 10 µm*) recorded at Chiang Mai City Hall from 2000–2012

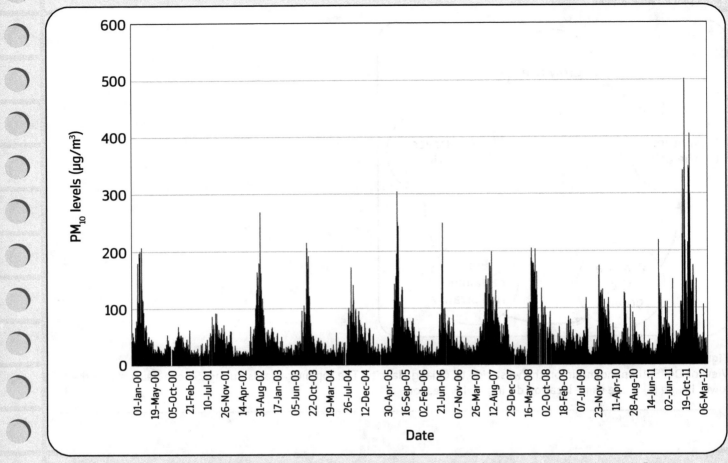

(Source: Adapted from Air Quality and Noise Management Bureau, Pollution Control Department (aqmthai.com), Ministry of Natural Resources and the Environment, Thailand (www.pcd.go.th))

Figure 10
Weekly average PM$_{2.5}$ levels recorded at four Chiang Mai sites, during 'Smoke Season' 2018

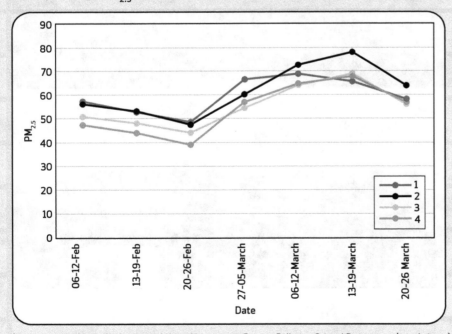

(Source: Adapted from Air Quality and Noise Management Bureau, Pollution Control Department (aqmthai.com), Ministry of Natural Resources and the Environment, Thailand (www.pcd.go.th))

Figure 11
Ecotourism Economic Cycle

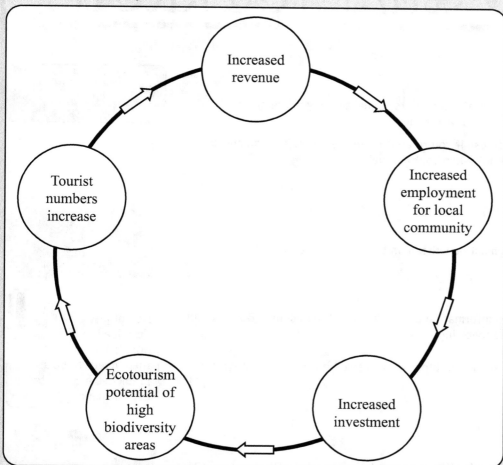

Figure 12
Tourism in Thailand fact file

Factfile Tourism in Thailand

In the past, tourism in Thailand was predominately based around international tourists and the capital of Bangkok as well as the Islands along the coast.

By 2018, the UN World Tourism Organization reported that domestic tourism was around 6 times greater than international tourism in Thailand.

Over the past decade, there as been a rise in ecotourism to the Highland Areas of Thailand including around Chiang Mai. Though there are also concerns that the word ecotourism is being use with little thought for impacts and affects by some tourism businesses.

Mae Kampong village, 55km north of Chiang Mae, has through the Thai Royal Projects scheme become a major ecotourism destination with many of the villages population of 500 offering homesteads.

The homesteads are operated as a cooperative with profits shared on projects to benefit the whole village.

Flight of the Gibbon, a zipline company operates a resort with over two hours of ziplines through the rainforest. Bringing tourists who may not have come just for the village and rainforest.

Profits from the company help support ecotourism and conservation within Mae Kampong.

TESTING WHAT YOU KNOW

Set B

The papers in this section have fewer tips and hints so make sure you are more confident with your revision before you tackle them.

Make sure you have extra paper to use if you run out of answer lines. You can check your answers in the back of the book when you're done!

Your Resource Booklet for this paper is on page 66.

Paper 1

- Answer **all** questions. Answers must be written on the answer lines provided.
- Set your timer for 1 hour.
- There are 35 marks available.
- The Resource Booklet provides information on Bhutan. Use the Resource Booklet and your own studies to answer the following.

Look at the topographical map. This shows mountains and lowlands.

Do not just describe the population pyramid, there needs to be a predictive element to get full marks.

1. Using Figures 1(a) to 1(c) **state <u>two</u>** types of terrestrial biomes present within Bhutan. **[1]**

    ....

    ....

2. Using Figure 2 and Figure 3, **identify <u>two</u>** population trends in Bhutan. **[2]**

    ....

    ....

    ....

    ....

    You need to look at both sets of information. The population pyramid can be used to predict what might happen in the future.

3. Using Figures 2 and 3, **identify <u>two</u>** reasons why Bhutan could be considered to be at Stage 3 of the demographic transition model. **[2]**

    ....

    ....

    ....

    ....

4. **Calculate** the rate of natural increase using Figure 2. **[1]**

ANSWER ANALYSIS

You must include % to gain the one mark. If you do not show the units you will not get any marks.

5. Using Figures 4(a), 4(b) and 6, **suggest <u>two</u>** ways in which Bhutan can further improve the development of renewable energy production to improve its energy independence and security. **[2]**

....

....

....

....

....

 Start by stating what energy independence and security are and how Bhutan currently fits into these ideas. Look at the maps and other information to see where else Bhutan could develop renewable energy, giving reasons why this might be successful.

6. Using Figure 6(b), **describe** the trend in the ecological footprint of Bhutan between 1990 and 2017. **[1]**

....

....

....

DESCRIBE
A describe question like this wants you to say what patterns (trend) you see in the data.

7. Using Figures 6(a) and 6(b), **outline** why a larger ecological footprint is seen after 2012. **[1]**

....

....

....

There is only one mark available, so do not write too much or spend too much time on this question.

8. Bhutan has been described as an ecocentric country. Using Figure 7, **outline <u>two</u>** reasons to support this description. **[2]**

....

....

....

....

....

Think about how an environmental value system can be developed. Link these factors to information provided about Bhutan. You need to provide evidence from the information that supports your answer.

9. Using Figures 1, 5, 8, 9 and 10, **explain** why Bhutan might have such a high number of large protected species such as the Bengal tiger, the snow leopard, the Asiatic black bear, the red panda and the Asian elephant. **[5]**

....

....

....

....

....

....

....

....

You need to explain the link between high numbers of large protected species and the measures that allow the protection.

ANSWER ANALYSIS

For your answer, think about what these big animals need. Larger animals need more space, less disturbance and less habitat fragmentation. The protected status of the forests means these requirements can be met.
The presence of large animals also indicates the good health of an ecosystem due to the amount of energy needed to support them.

10. Using information from Figure 4(b) and 6, **outline one** environmental impact of developing the proposed hydroelectric stations. **[1]**

....

....

....

....

....

OUTLINE
Outline needs to include the cause and the effect.

Hydroelectric power produces power through the damming of water. While this is a very effective way to generate power in a place that has a lot of freshwater, it can also have detrimental impacts on the environment. Dams are normally built in steep sided valleys and they require construction over a long period of time often in areas that are isolated from urban development.

11. Using information from Figure 4(b) and 6, **explain** the impacts of developing additional hydroelectric stations on biodiversity. **[3]**

....

....

....

....

....

....

....

What is the impact on trees of developing additional hydroelectric stations? What about the impact on fish?

12. Before any new hydroelectric scheme is implemented an EIA (Environmental Impact Assessment) needs to be carried out. **Describe** the purpose of an EIA. **[1]**

....

....

....

....

ANSWER ANALYSIS

Why would an EIA be needed and what is it for? This question is related to your knowledge of EIAs, rather than information you will find in the Resource Booklet.

13. **Suggest** how a team of ecologists could measure changes in biodiversity in the forest around a hydroelectric dam over a period of 10 years after its construction. **[3]**

....

....

....

....

....

....

....

You may not have studied exactly the same case study before, but use what you know about ways of measuring biodiversity to make a suggestion. This is not a detailed IA method.

To look at change, there must be data from before the construction as well as after to allow a comparison to be made.

14. Using all the information provided, **identify <u>two</u>** possible positive impacts of increasing economic development over environmental conservation in Bhutan.

[2]

....

....

....

....

 There will be more than two possible positive impacts.

Read the question carefully.

15. Using all the information provided, **identify <u>two</u>** possible negative impacts of increasing economic development over environmental conservation in Bhutan.

[2]

....

....

....

....

There will be more than two possible negative impacts.

16. Using the information in Figure 10 and the other resources, **evaluate** the sustainability of Bhutan's tourism policy.

[6]

....

....

....

....

....

....

....

....

....

....

....

....

....

....

Evaluate answers need to provide **both** advantages and disadvantages.

!

ANSWER ANALYSIS

Think of how tourism can have negative impacts on areas. Are these relevant in Bhutan due to the low number of tourists? How has limiting the number of tourists protected the country? What are the negative sides of limiting tourism? A one-sided argument will not allow you to access the highest level grades.

Set B: Paper 2

- Set your timer for **2 hours**
- Section A – answer **all** the questions.
- Section B – answer **two** questions.
- The maximum mark for this examination paper is **65 marks**.

Section A

1. **Figure 1: Percentage cooling energy consumption between 1990 and 2016**

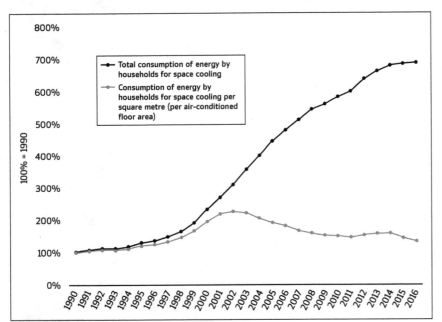

(Source: The European Environment Agency (CC BY 2.5 DK))

(a) **State** the year that the difference between consumption was first greater than 100%. **[1]**

....

(b) **Calculate** the change in total consumption from 1990 to 2016. **[1]**

(c) **Identify** **one** potential reason for the exponential rise in the total amount of energy used in the cooling of houses compared to the energy used per square metre. **[1]**

....

....

(d) **Describe** how CFCs impact on the stratospheric ozone concentration. **[1]**

....

....

....

....

If you are asked to calculate something make sure that you write out your calculation. You must use the units to get the mark. If you forget the units you will get zero marks for this question.

STATE
Recognize and state the year only.

Use a ruler to get the most accurate values you can.

Spend a little time looking at both data sets shown on the graph. How are they related to each other?

(e) **Outline two** effects on human tissue of ultraviolet radiation. **[2]**

....

....

....

....

....

(f) **Identify** why even following the success of the Montreal Protocol, CFCs are still being released into the atmosphere. **[1]**

....

....

2. **Figure 2: Global map showing the location of areas of physical and economic water scarcity**

Little or no water scarcity
Physical water scarcity
Approaching physical water scarcity
Economic water scarcity
Not estimated

(Source: World Water Development Report 4. World Water Assessment Programme (WWAP), March 2012)

(a) Ground water is considered Renewable Natural Capital. **Define** Renewable Natural Capital. **[1]**

....

....

(b) **Distinguish** between economic water scarcity and physical water scarcity, providing a named location for each. **[2]**

....

....

....

....

(c) Using a named example, **outline** how human impacts have resulted in the overuse, pollution and depletion of a freshwater system. **[3]**

....

....

....

....

....

....

....

IDENTIFY

There are a number of possible answers and you need to select one of them.

Maps are often used to show the distribution of things. You need to have a basic understanding of where different countries are located to be able to clearly interpret findings and patterns.

ANSWER ANALYSIS

To get the full mark for each point you need to include the definition and a named example.

DISTINGUISH

Show a clear difference between two or more ideas, concepts or definitions.

(d) **Outline** <u>two</u> ways to enhance the supply of water in countries that experience water scarcity. **[2]**

....

....

....

....

(e) **Draw** a flow diagram to illustrate the flows of inorganic fertilizer, eutrophication and nitrate enrichment of aquatic systems from agricultural systems. **[2]**

ANSWER ANALYSIS

You need to be careful with this question that you are using examples that **enhance** rather than conserve. For example, using grey water would be a conservation method not an enhancement.

The question is asking about transfers and transformations. What are the conventions used when constructing this type of diagram?

Use boxes as stores and arrows to show flows, similar to a simple systems diagram. You do not need to draw a picture.

3. **Figure 3: Percentage of bee colonies lost within the United States from January to March, 2017**

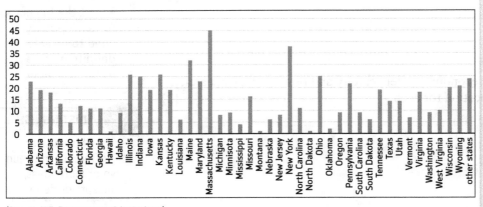

(Source: U.S. Department of Agriculture)

(a) Bees have been called indicators of environmental change. **Define** indicator species. **[1]**

....

....

(b) **Outline** <u>one</u> reason why bees are so important within an ecosystem. **[1]**

....

....

....

ANSWER ANALYSIS

Learn your definitions for easy recall.

(c) Some states have lost more than one quarter of their bee colonies. **Outline two** potential impacts of a further loss of bees within these states. [2]

....

....

....

....

(d) **Suggest** a potential solution to address the reduction in pollinators in agricultural systems. [4]

....

....

....

....

....

....

....

....

Use a ruler to determine which countries have values over 25%.

Think about the wider impacts and remember that all parts of a system are interrelated.

As you are being asked to suggest something it does not have to be a project that is currently happening. Make it a feasible project and present a little information on the potential issues that may occur.

Section B

Choose any **two** from the list of three questions. (Remember, in the real exam you will choose two from a list of four.)

4. (a) **Distinguish** between storages and processes that are part of the carbon cycle.
[4]

> Think about having a quick look at these questions during your 5 minutes reading time. This will help you decide which you are going to answer. You could even spend a couple of minutes noting down a quick plan once your 5 minutes thinking time is finished.

> **!** Make sure that you are giving places where carbon is kept and ways in which it is built up or broken down.

ANSWER ANALYSIS

To get full marks here you need to give a brief answer for each area. Link it to the carbon cycle, e.g. 'Decomposition is a process where dead plants and animals are broken down, releasing the stored carbon'.

(b) **Explain** the ways in which human activity has impacted on the carbon cycle.
[7]

> **!** Do not include any natural impacts on the cycle as these will not get you any marks.

> **...** **EXPLAIN**
> Give a detailed explanation with reasons and causes. Some of the marks will be for describing the impacts but to reach full marks you will need to link those to reasons. You will need to explain **how** human activity has impacted on the carbon cycle.

> Start by giving a brief description of the carbon cycle and the links to human activities. Then present either three cases with limited detail or two with more detail.

> **!** Notice that you are not asked for positive or negative impacts so you can present either side, or both.

(c) With reference to a named country, **discuss** the advantages and disadvantages of renewable energy on reducing impacts on the carbon cycle. **[9]**

....

....

....

....

....

....

....

....

....

....

....

....

....

....

....

....

5. (a) **Distinguish** between in-situ and ex-situ conservation. **[4]**

....

....

....

....

....

....

....

(b) Both in-situ and ex-situ methods are used in maintaining biodiversity.

Explain for a named ecosystem what procedures could be used to compare the effect of in-situ conservation methods on biodiversity of that ecosystem. **[7]**

....

....

....

....

....

....

....

Begin by giving a definition of renewable energy and its link to human impacts on the carbon cycle. Then give a number of examples and a balanced conclusion stating the current level of use of renewable energies.

ANSWER ANALYSIS

If you do not include the name, location and type of project in your answer you will not be able to access the highest grades for this question.

ANSWER ANALYSIS

You can use any example that you know in your answer but you need to be able to give enough detail to show how it positively and negatively links to the carbon cycle to get the marks.

DISTINGUISH

The differences need to be clear.

Read this question carefully. Despite mentioning in-situ and ex-situ conservation in the initial statement, the question is only related to in-situ conservation.

Answer lines continue on next page

....

....

....

....

....

....

(c) **Evaluate** the success of a named protected area in reducing the impacts of human activity on biodiversity. **[9]**

....

....

....

....

....

....

....

....

....

....

....

....

....

....

....

....

....

....

ANSWER ANALYSIS

Make sure you are using ES&S terminology correctly throughout your writing in order to access the highest marks.

ANSWER ANALYSIS

For full marks you need a short introductory statement about biodiversity loss and the causes and to give the positive and negative aspects of a number of strategies (with specific examples). End with what you think about the situation. A one-sided answer will reduce the number of marks your answer is able to get.

6. (a) **Identify four** ways in which individual households can reduce the amount of waste they send to landfill. **[4]**

....

....

....

....

....

....

....

ANSWER ANALYSIS

You are being asked to just give a list of ways here, so you do not need to give any detail. For full marks you need to give four things, making sure they are clearly different strategies. Do not split recycling into recycling glass and recycling paper expecting to get a mark for each point.

(b) **Explain** how recycling programs can be used as a reduction strategy for solid domestic waste. [7]

There are many aspects that you could talk about. Just make sure that you present the strengths and weaknesses of the strategy and give a conclusion about how good recycling is as a waste management strategy.

Make sure that you know your command terms so that you can distinguish between what is being asked of you in different questions.

Include a case study to demonstrate how effective recycling programs can be.

'Upcycling' is the process of taking old, used items and creating some other item with it. E.g. turning old t-shirts into shopping bags, making items from used plastic bags.

(c) There are three levels of pollution management relating to different stages of activity. **Discuss** how each level can be applied to the problems of solid domestic waste. [9]

Each of the three pollution management levels needs to be addressed, but make sure that you give a short description of the three levels first.

ANSWER ANALYSIS

A question like this is not about having a set number of examples. It is about the way that you integrate all of the aspects together. To access the 7–9 mark band you need to make sure that your answer is thorough, balanced and supported by robust examples that are critically reflected upon.

RESOURCE BOOKLET

Set B: Paper 1

This Resource Booklet is for use with Set B, Paper 1 on page 54.

Figure 1(a)
Location of Bhutan within Asia

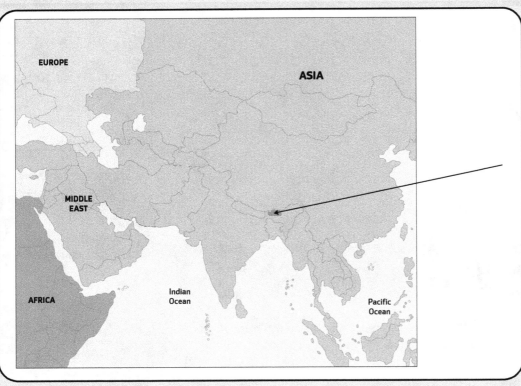

Figure 1(b)
Topographical map of Bhutan

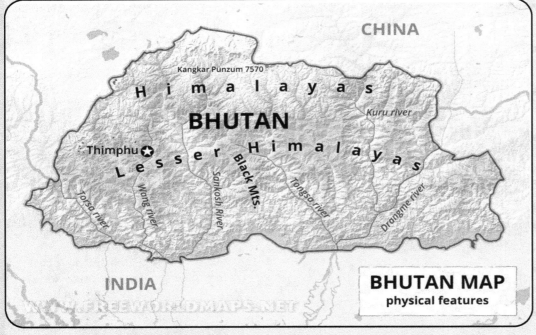

(Source: www.freeworldmaps.net/asia/bhutan/map.html)

Figure 1(c)
Land Cover in Bhutan

Class	Sub-Class	Area (ha)
Forests	Conifer Forest	983,240
	Broadleaf Forest	1,720,295
Shrub Vegetation	-	419,128
Grasslands	-	157,238
Agricultural Land		112,138
Built-Up Urban Area	-	6,194
Non-Built Up Urban Area	-	330
Permanent Snow Cover	-	299,339
Bare Areas with no vegetation		130,829
Water Bodies	Lakes	4,751
	Reservoirs	131
	Rivers	22,563
Marshy Areas	-	319
Degraded Areas	-	20,602
Total Area		3,877,097

(Source: Adapted from data in The Project for Formulation of Comprehensive Development Plan for Bhutan 2030 Final Report. Ministry of Works and Human Settlement (MoWHS) and Japan International Cooperation Agency (JICA), https://www.gnhc.gov.bt/en/wp-content/uploads/2019/11/EIJR19080-CNDP-Vol1_Summary.pdf)

Figure 2
Fact file on the population of Bhutan

- Area of the country: 38,394 sq km.
- Population: 766,397 people in 2018.
- Ethnicity: Ngalop 50%, ethnic Nepali 35%, indigenous or migrant tribes 15%.
- Main religion: Bhuddism.
- Population growth: 1.05%.
- Birth rate: 17 births/1000 population.
- Death rate: 6.4 deaths/1000 population.
- Urbanization: 2.98% annual change.

(Source: Central Intelligence Agency)

Figure 3
Population pyramid for Bhutan from 2017

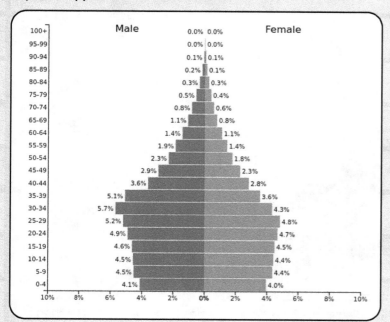

(Source: www.populationpyramid.net/bhutan/2017)

Figure 4(a)
Classification of wind power potential in Bhutan

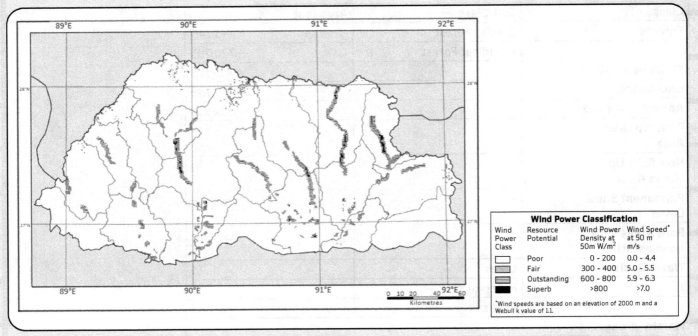

(Source: Adapted from an original map produced by NREL together with USAID and Department of Energy, Ministry of Economic Affairs, Royal Government of Bhutan, www.nrel.gov/international/ra_bhutan.html)

Figure 4(b)
Locations of Bhutan's hydroelectric power stations

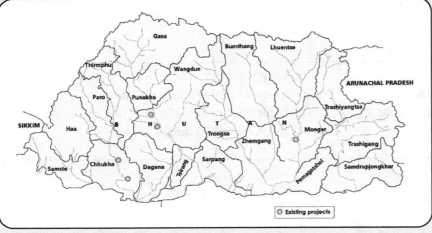

(Source: Central Electricity Authority India)

Figure 5
Percentage tree cover in different Dzongkhags within Bhutan

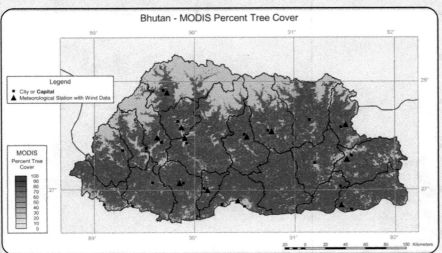

(Source: Adapted from an original map produced by NREL together with USAID and Department of Energy, Ministry of Economic Affairs, Royal Government of Bhutan, www.nrel.gov/international/ra_bhutan.html)

Figure 6(a)
Fact file on Bhutan's energy use

- 100% of the urban population but only 53% of rural communities had electricity in 2012.
- The government subsidizes power to rural communities.
- An estimated 7.883 billion kWh of power was generated in Bhutan in 2016.
- Bhutan used an estimated 2.184 billion kWh of power in 2016.
- An estimated 5.763 billion kWh of power were exported from Bhutan in 2016.
- Currently only 5% of the country's hydropower resources are being accessed.
- Bhutan has been exporting hydroelectric power to India since 1974, accounting for 22% of Bhutan's GDP in 2002.
- Approximately 70% of Bhutan's hydroelectricity goes to India.
- The export of hydroelectricity is now the single largest source of income for Bhutan.
- To reduce reliance on hydroelectricity, other renewable sources such as solar, biogas and wind are being developed.

(Source: information collected from Central Intelligence Agency and IB extended essay interviews)

Figure 6(b)
Ecological footprint of Bhutan 1990–2017

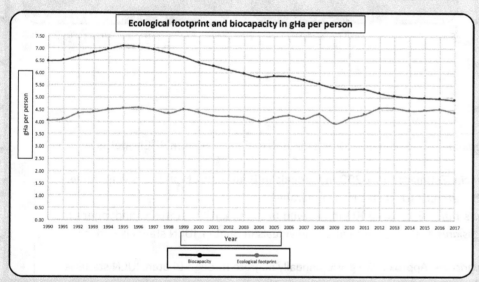

(Source: Data extracted from https://data.footprintnetwork.org/)

Figure 7
Fact file on Bhutan's Gross National Happiness

- Gross National Happiness is as, if not more, important as Gross Domestic Product.
- GNP is a sustainable and holistic approach to progress.
- GNP focuses on four main pillars:
  1. Sustainable development
  2. Environmental conservation
  3. Cultural preservation
  4. Good governance
- Bhutan mandates that at least 60% of the country needs to be forested; 72% of the country is currently forested.
- Around 50% of the country is environmentally protected.
- Bhutan is known to be the only carbon negative country in the world.
- Television broadcasting in Bhutan started in 1999 and cable TV was available by 2012.
- Telephone communication was poor in rural areas until cellular service became widely available in 2012.

Figure 8

(a) Red panda

(Source: Unsplash.com, Dušan Smetana)

(b) Black-necked crane

(Source: Shutterstock, Wang LiQiang)

(c) Snow leopard

(Source: Tambako the Jaguar, Djamila lying bravely but looking seriously, flickr.com, (CC BY-ND 2.0))

(d) Asian elephant

(Source: Bernard DUPONT, Asian Elephant (Elephas maximus), flickr.com, (CC BY-SA 2.0))

(e) Whited-bellied heron

(Source: Wikicommons, Rohit Naniwadekar)

(f) Asiatic black bear

(Source: Jean-Pierre Dalbéra, Ours à collier (Zoo de Berlin), flickr.com, (CC BY 2.0))

Figure 9
Fact file on Bhutan's biodiversity

- Asian elephant – *Elephas maximus*. Seasonal migration between India and Bhutan. IUCN status is endangered.
- Snow leopard – *Panthera uncia*. Estimated 100-200 individuals in Bhutan. IUCN status is vulnerable.
- Asiatic black bear – *Ursus thibetanus*. IUCN status is vulnerable.
- Bengal tiger – *Panthera tigris tigris*. Estimated 103 in the wild in Bhutan. IUCN status is endangered.
- White-bellied heron – *Ardia insignis*. Estimated around 30 individuals in Bhutan. IUCN status is critically endangered.
- Black-necked crane – *Grus nigricollis*. Approximately 500 annually migrate through Bhutan. IUCN status is vulnerable.
- Red panda – *Ailurus fulgrens* estimated at around 2500 remaining in the wild. IUCN status is endangered.

(Source: WWF, www.wwfbhutan.org.bt/_what_we_do/wildlife/species/ (CC BY-SA 3.0))

Figure 10
Bhutan's low impact tourism policy

- Tourism began in 1974.
- High value and low impact.
- Tourism is dictated by the principles of sustainability to make it an eco-friendly and culturally and economically acceptable activity.
- A daily tourist tariff is applied that covers the cost of guides, accommodation, food and transportation. You can spend more than this on your visit, but not less.
- Reputable guides must be used for all tourist activities, thus limiting places where tourists can visit and stay.
- 35% of the tourist tariff helps to provide free medical care and schooling for all Bhutanese people.

(Source: Adapted from: www.bhutantravelbureau.com/getting-to-bhutan/bhutan-tourism-policy/)

Set C

Paper 1

- Answer **all** questions. Answers must be written on the answer lines provided.
- Set your timer for 1 hour.
- There are 35 marks available.
- The Resource Booklet provides information on Newfoundland. Use the Resource Booklet and your own studies to answer the following.

Your Resource Booklet for this paper is on page 82.

1. Using Figure 1(a), 1(b), 1(c) and 2, **outline one** reason for Newfoundland having biologically poor species numbers. **[1]**

2. Using Figures 4 and your own knowledge, **identify two** types of population interactions in the marine ecosystem. **[2]**

   Population interactions have to do with more than just feeding.

3. With reference to Figure 2, **identify two** economic activities in Newfoundland. **[2]**

4. Using Figure 2, **calculate** area of Newfoundland that is covered by forest. **[1]**

   Think about the conventions for calculation questions.

5. Using Figure 5(b), **draw** a food chain that includes four trophic levels. **[2]**

6. **Outline two** possible reasons why most people are found along the coast of the island. **[2]**

7. There are advantages and disadvantages from the use of factory fishing in the fish industry in Newfoundland.

   (a) **Suggest <u>two</u>** advantages. **[2]**

....

....

....

....

   (b) **Suggest <u>two</u>** disadvantages. **[2]**

....

....

....

....

8. Using Figure 7:

   (a) **Describe** the trend in Newfoundland's population between 2000 and 2017. **[2]**

....

....

....

   (b) **Suggest <u>two</u>** reasons for the trend in population between 2007–2016. **[2]**

....

....

....

   (c) **Explain** why it is difficult to calculate the carrying capacity for human populations. **[2]**

....

....

....

....

9. Using Figure 2 and Figure 8(a), **identify <u>one</u>** form of natural capital and **<u>one</u>** form of natural income. **[1]**

....

....

10. Using Figure 1(c), Figure 3(a) and Figure 8(b), **explain** the percentage share of GDP that comes from agriculture in Newfoundland. **[3]**

....

....

....

....

....

....

ANSWER ANALYSIS

'Suggest' questions allow you to think of possibilities that may not necessarily be in the Resource Booklet.

ANSWER ANALYSIS

For 'describe a trend' questions, you need to say exactly what you see and include numbers.

DESCRIBE

To give a detailed account.

ANSWER ANALYSIS

Give at least **one** reason for the increase and **one** reason for the decrease.

Natural capital is a natural resource that gives income.

ANSWER ANALYSIS

Make sure you read what the question requires... It's easy to be caught out by thinking a 1 mark question only needs 1 answer.

ANSWER ANALYSIS

You need to give details to get the full 3 marks, so you need to say what the GDP is and that it is low, then give at least two reasons why.

11. Using Figure 9(a) and 9(b) and the other resources, **identify two** effects of mining iron ore on the environment in Newfoundland. [2]

....

....

....

....

12. *Braya longii* is an endemic species found only in a few small populations on Newfoundland. **Explain** how plate tectonics contributes to the evolution of endemic species. [3]

....

....

....

....

....

....

....

....

....

 EXPLAIN

Remember, you need to give a detailed answer with supporting reasons.

 What does the term 'endemic' mean?

Use an example of an area that clearly supports your answer.

13. Both species of *Braya* have low population numbers due to quarrying.

To what extent can focusing on habitat conservation aid the conservation of individual species? [6]

....

....

....

....

....

....

....

....

....

....

....

....

....

....

A 'To what extent...' question needs to examine both strengths and weakness . You do not need to include all of the possibilities for and against but you must have at least 5 points in total from both for and against to be awarded 5 marks.

 ANSWER ANALYSIS

The sixth mark comes from drawing a conclusion. The conclusion needs to be valid and balance between both strengths and weaknesses within the arguments. It needs to have explicit details that give evidence to the conclusion.

Set C: Paper 2

- Set your timer for **2 hours**
- Section A – answer **all** the questions.
- Section B – answer **two** questions.
- The maximum mark for this examination paper is **65 marks**.

Section A

1. (a) **State <u>one</u>** component of soil system storage. **[1]**

....

....

....

....

> Give a summary in the form of a sentence. Avoid one-word answers for this question.

(b) **Outline <u>two</u>** factors that determine a society's food choices. **[2]**

....

....

....

....

Figure 1: Energy required to produce one pound of food

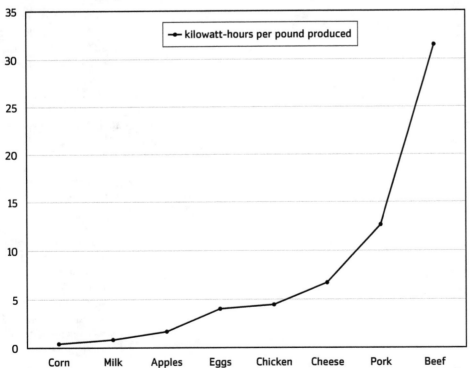

(Source: Adapted from truecostblog.com/tag/food-energy/ and www.treehugger.com/green-food/energy-required-to-produce-a-pound-of-food.html)

> Make sure you refer to the diagram when you are answering the question.

(c) **Outline two** reasons why it is more energy efficient for humans to feed on corn than on beef. [2]

....

....

....

....

....

(d) **Distinguish** between the inputs of factors of scale and labour in terrestrial food production between MEDCs and LEDCs. [2]

....

....

....

....

....

....

....

> Give a detailed account that includes reasons. Give at least two corresponding differences.

> Using a table can make sure you include the correct number of differences.

2. **Figure 2**: Soil readings from three different sites in a temperate woodland

Site 1   Site 2   Site 3

pH   Moisture content   Dissolved nitrates   Microbe diversity

(a) **State** whether the above factors in Figure 2 are biotic or abiotic. [1]

....

(b) **Describe** how you could measure the moisture content in the soil at the three sites. [2]

....

....

....

....

> Give a technique that is easy to repeat in each area to allow for comparisons.

Figure 3: Food web

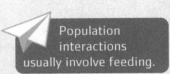

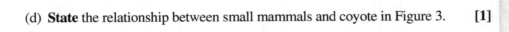

(c) **Draw** a food chain from the food web shown in Figure 3 that includes trophic level 4. **[1]**

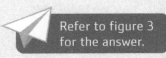

(d) **State** the relationship between small mammals and coyote in Figure 3. **[1]**

....

....

(e) **Outline** how the relationship might benefit both the small mammals and the coyote populations. **[2]**

....

....

....

....

(f) **Outline** why the bird of prey can be considered in two trophic levels. **[2]**

....

....

....

....

....

Use the reading time to think about this diagram. It has a lot of detail.

Population interactions usually involve feeding.

Lines on food chains **MUST** have arrows indicating energy flow.

Remember that a food chain is one line out of the food web.

Refer to figure 3 for the answer.

3. (a) **Define** species diversity. [1]

....

....

Definitions must be precise.

This question addresses AO1 only, as a definition is asked for.

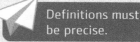

(b) **Outline**, for a named species, <u>one</u> human cause of extinction. [2]

....

....

....

....

Think of HIPPO – an acronym to help you remember the human activities that result in extinction of species.
For two marks you need only brief information.

(c) **Distinguish** between Flagship species and Keystone species as used in conservation. [2]

....

....

....

....

Give a clear difference between each.

Before starting, underline the key words here: 'flagship species' and 'keystone species'.

(d) Conservation approaches include habitat conservation.

Explain factors that should be considered in designing protected areas. [4]

....

....

....

....

....

....

Think of SLOSS: **Single Large Or Several Small.** This acronym can help you remember that a single large area is preferable.

Section B

Choose any **two** questions from the list of three. (Remember, in the real exam you will choose two from a list of four.)

4. (a) **Outline** the role of the greenhouse effect in regulating temperature on Earth. **[4]**

....

....

....

....

....

....

....

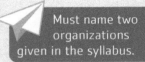

Only choose two questions. You should not attempt to answer all of them. But, after you have finished the practice paper, go back and try the question you did not choose for more practice.

(b) "The international community has been successfully working together to preserve the environment for more than 30 years."

Evaluate this statement with reference to both national and international organizations in reducing the emission of ozone depleting substances. **[7]**

....

....

....

....

....

....

....

....

....

....

....

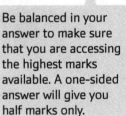

Must name two organizations given in the syllabus.

ANSWER ANALYSIS

Be balanced in your answer to make sure that you are accessing the highest marks available. A one-sided answer will give you half marks only.

(c) The Arctic is warming much faster than the rest of the planet and the loss of reflective ice contributes somewhere between 30–50% of Earth's global heating.

With reference to the statement **discuss** the role of feedback mechanisms associated with a change in mean global temperature. **[9]**

....

....

....

....

....

....

....

....

....

....

....

DISCUSS

To put all of the aspects together and present the individual and cumulative impacts. Think about how a change in one aspect can have a ripple effect of changes throughout the area.

ANSWER ANALYSIS

A 9-mark question should take you around 12–13 minutes to complete. It's a detailed answer.

Answer lines continue on next page

....

....

....

....

....

....

DISCUSS

To give a clear and balanced view that includes different factors and evidence to support both sides of the argument.

5. (a) With reference to named examples, **distinguish** between the terms succession and zonation. **[4]**

How are these two types of habitat distribution different?

....

....

....

....

....

....

....

(b) **Explain** how plate tectonic activity has influenced evolution and biodiversity. **[7]**

This question addresses AO3, as you must use given information and reasons in your answer.

....

....

....

....

....

Do not just present one idea for this many marks.

....

....

....

....

....

....

(c) **To what extent** have different approaches to protecting biodiversity from the threats of human activity for a named protected area been successful? **[9]**

Make a quick plan to ensure you cover all areas enough.

....

....

....

....

....

Answer lines continue on next page

....

....

....

....

....

....

....

....

....

....

....

....

It is always best to use any of the case studies given in the syllabus.

6. (a) The hydrological cycle is a system. With reference to processes occurring within it:

    (i)   **Identify <u>two</u>** transformations of matter.   **[2]**

....

....

....

    (ii)   **Identify <u>two</u>** stores.   **[2]**

....

....

  (b) **Explain** how human activity impacts the hydrological cycle using a systems diagram.   **[7]**

Stores are shown as enclosed boxes. Processes are flows so they must have arrows.

Plan this out to make sure that you include all areas of the cycle.

(c) **Discuss** the strategies that can be used to meet an increasing demand for fresh water with reference to Environmental Value System. **[9]**

....

....

....

....

....

....

....

....

....

....

....

....

....

....

....

....

....

....

....

Give a range of factors or arguments. Have a conclusion that is clear and supported by evidence.

Make sure that your answer is coherent, as if it does not make sense or flow logically it is hard for the examiner to follow your ideas, which may result in lower marks. Make a quick plan before you start.

When you are asked to discuss, you are not able to reach the higher level marks if you only present one side of the argument.

Give at least 3 strategies, then evaluate each by giving its strengths and weaknesses.

RESOURCE BOOKLET

Set C: Paper 1

This Resource Booklet is for use with Set C, Paper 1 on page 71.

Figure 1(a)
World map showing the location of Canada

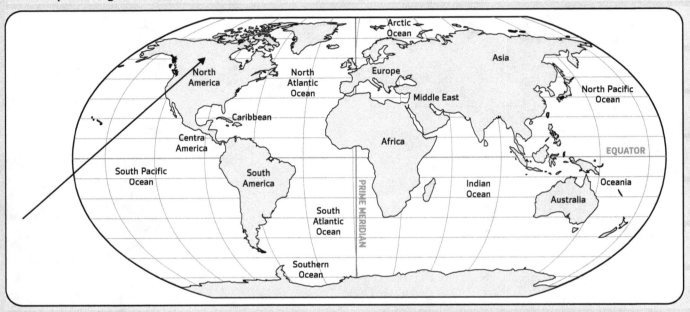

Figure 1(b)
Map showing the location of Newfoundland in Canada

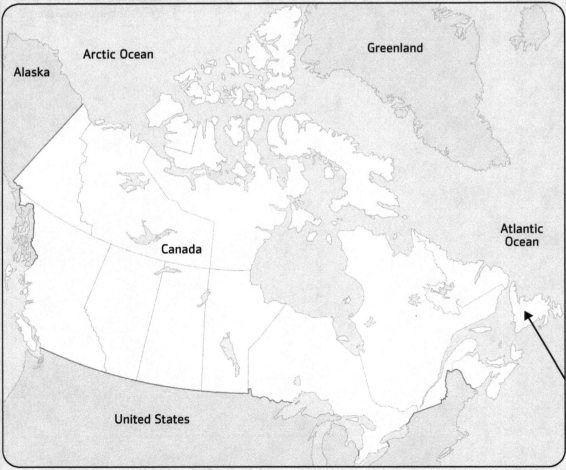

Figure 1(c)
Solar spreading at the poles

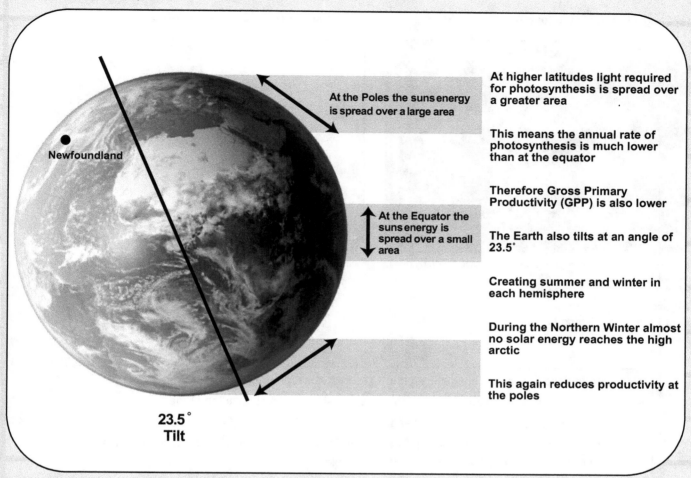

At higher latitudes light required for photosynthesis is spread over a greater area

This means the annual rate of photosynthesis is much lower than at the equator

Therefore Gross Primary Productivity (GPP) is also lower

The Earth also tilts at an angle of 23.5°

Creating summer and winter in each hemisphere

During the Northern Winter almost no solar energy reaches the high arctic

This again reduces productivity at the poles

At the Poles the suns energy is spread over a large area

At the Equator the suns energy is spread over a small area

Newfoundland

**23.5°
Tilt**

(Source: N Gardner, Four Corners Education, fourcornerseducation.net)

Figure 2
General fact file on Newfoundland

- The easternmost province in Canada.
- Land area is 108,860 km².
- Fishing and aquaculture account for 3% of the annual gross GDP. Cod is the staple of the fishing economy.
- Mining accounts for 4% of the province's annual GDP with iron ore being the most important mineral.
- Most of the soil is unsuitable for farming and agriculture accounts for less than 1% of the annual GDP.
- 60% of Newfoundland is forested.
- Population distribution is uneven with most people along the coast of the island.
- Gros Marine National Park, formed mainly because of plate tectonics, is a world heritage site.
- The island is biologically poor in species number but rich in biomass.
- Long's braya (*Braya longii*) and Fernald's braya (*B. fernaldii*) are rare herbaceous plant species that are endemic to Newfoundland.
- The braya poluplation is low due to quarrying.

(Source: https://www.gov.nl.ca/publicat/royalcomm/research)

Figure 3
braya plants

(a) Long's braya (*Braya longii*)

(Source: Gene Herzberg, Braya longii, Long's braya, Sandy Cove, flickr.com (CC BY 2.0))

(b) Fernald's braya (*Braya fernaldii*)

(Source: Mark R. Tsang, 09 13 06_VikingTrail I_7568 Fernald's Braya (Rare), flickr.com)

Figure 4
Fact file on cod fishing in Newfoundland

- Newfoundland is a large Canadian island off the east coast of the North American mainland.
- Cod fishing in Newfoundland was carried out at subsistence level for a long time.
- Factory fishing using super-trawlers began in 1951.
- The number of cod caught peaked in 1968 with 810,000 tons caught. This was three times more than the number of cod caught per annum before the super-trawlers.
- Overfishing led to the collapse of fisheries in 1992.
- By 1993, six cod populations had collapsed leading to a ban on fishing. Spawning biomass had decreased by 75% in all stocks.
- The cod had not returned even after a 10-year ban on fishing.
- Slow recovery was attributed to a change in the local ecosystem; the presence of capelin, a small forage fish used to provide food for the cod but which was now eating the juvenile cod.
- By 2011, the fisheries were returning to their abundance.

(Source: https://en.wikipedia.org/wiki/Cod_fishing_in_Newfoundland)

Figure 5(a)
Some of the species found in the North Atlantic

Atlantic cod

(Source: Krasowit, shutterstock.com)

Capelin

(Source: Sergey Goruppa, shutterstock.com)

Atlantic Deep-Sea Red crab

(Source: cropped from rawpixel.com)

Atlantic Northern Shrimp

(Source: K321, shutterstock.com)

Figure 5(b)
Terrestrial food web for an area of Newfoundland

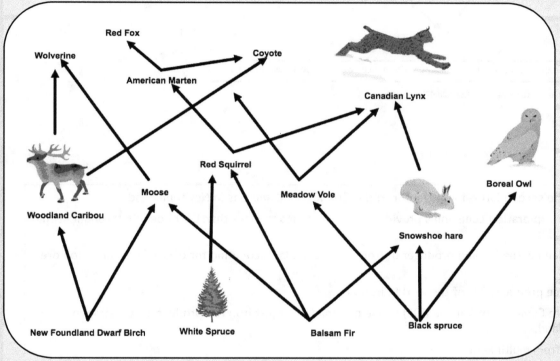

(Source: N Gardner, Four Corners Education, fourcornerseducation.net)

Figure 6
Collapse of Atlantic cod stocks (East Coast of Newfoundland), 1992

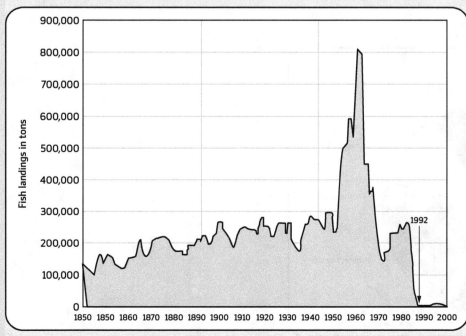

(Source: SatCan, www150.statcan.gc.ca/t1/tbl1/en/tv.action?pid=1710000501)

Figure 7
Population estimates for Newfoundland and Labrador, Canada, from 2000–2017.

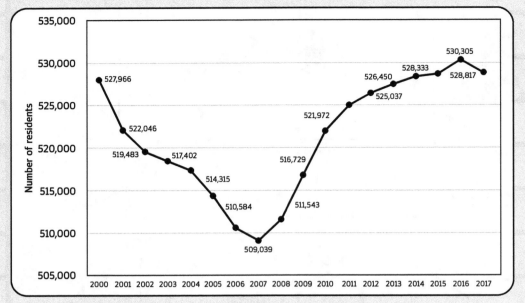

(Source: SatCan, www150.statcan.gc.ca/t1/tbl1/en/tv.action?pid=1710000501)

Figure 8(a)
Fact file on mining

- Mining is the second largest contributor to the GDP after Oil and Gas in Newfoundland.
- Mining and exploration companies provide high-paying jobs to more than 8000 people throughout the province.
- Newfoundland is the leading producer of iron ore in Canada, accounting for 63% of Canadian iron ore production.
- The province produces 2% of the world's iron ore.
- The Iron Ore Company of Canada (IOC) is one of Canada's largest iron ore producers operating in Newfoundland.
- IOC mines from multiple open pits.
- IOC has registered a different company for environmental assessment to support its ongoing operations.

Figure 8(b)
Relative share of GDP in Newfoundland

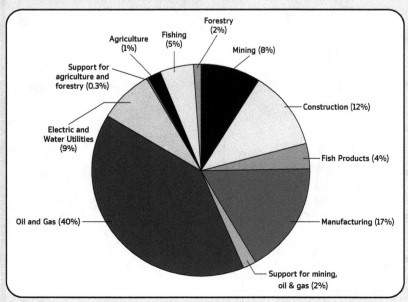

(Source: Newfoundland Statistics Agency and Statistics Canada)

Figure 9(a)
Open pit mining

(Source: iStock, SergeyZavalnyuk)

Figure 9(b)
Voisey's Bay Ovoid Pit

(Source: Ariontor, pixabay.com)

Set D

This set of papers has no additional help in the margin. There is space to write notes so you can plan what you are going to write if needed.

Paper 1

Your Resource Booklet for this paper is on page 100.

- Answer **all** questions. Answers must be written on the answer lines provided.

- Set your timer for 1 hour.

- There are 35 marks available.

- The Resource Booklet provides information on Venezuela. Use the Resource Booklet and your own studies to answer the following.

1. **State** the type of biome found in Venezuela. **[1]**

2. (a) Using Figure 2(b), **outline two** reasons for differences between the age-gender pyramids of 1990 and 2015 for Venezuela. **[2]**

   (b) Using Figure 2(b), **identify** the stage of demographic transition model (DTM) represented by the pyramid for 2015. **[1]**

3. Figure 3(a) shows hydroelectric power consumption in Venezuela over 26 years.

   (a) **Outline two** advantages of hydroelectric power. **[2]**

      1

      2

   (b) Figure 3(b) shows a relatively constant and low level of usage of renewable energy sources in Venezuela between 1990 and 2015.

      Suggest two reasons for this. **[2]**

      1

      2

NOTES

(c) With reference to Figure 4(a), **suggest <u>one</u>** other economic benefit of the Guri Dam other than hydroelectric power generation. **[1]**

....

....

(d) With reference to Figure 4(b), **suggest <u>one</u>** reason for the change in water level in the Guri Dam with time. **[1]**

....

....

4. (a) With reference to Figure 5(c), **identify <u>two</u>** reasons why freshwater supply is limited in Venezuela. **[2]**

....

....

....

....

....

....

(b) With reference to Figure 5(a), Figure 5(b) and Figure 5(c), **explain** how contamination with agricultural waste can lead to eutrophication. **[3]**

....

....

....

....

....

....

....

(c) With reference to Figure 5(a) and Figure 5(b), **outline** how Venezuela residents face an increased risk of Zika infection. **[3]**

....

....

....

....

....

(d) Mosquitoes also transmit the malaria parasite (*Plasmodium falciparum*) to humans. **Define** parasitism. **[1]**

....

....

NOTES

5. Venezuela is one of the ten countries with the highest biodiversity on Earth.

(a) With reference to Figure 1(d) and Figure 6(a), **identify two** reasons for the high biodiversity in Venezuela's rainforests. **[2]**

....

....

....

....

(b) With reference to Figure 6(b), **calculate** the total percentage of land in Venezuela that has over 30% forest cover. **[1]**

(c) Ecosystems provide natural income in terms of products and services. **Identify one** product and **one** service that may be obtained from Venezuela's rainforests. **[2]**

....

....

....

....

6. With reference to Figure 7(a), **suggest two** reasons why the giant otter was classified as endangered. **[2]**

....

....

....

....

7. With reference to the species shown in Figure 7, **explain** the reasons for preserving the biodiversity of Venezuela's rainforests. **[4]**

....

....

....

....

....

....

....

....

NOTES

8. In the year 2000, the UN published the Millennium Development Goals (MDGs). Goal 5 of the MDGs targeted 'improved maternal health' and Goal 6 targeted 'reversing and stopping malaria incidence'.

With reference to Figure 8, **to what extent** had these goals been achieved in Venezuela by 2015?   [5]

NOTES

Set D: Paper 2

- Set your timer for **2 hours**
- Section A – answer **all** the questions.
- Section B – answer **two** questions.
- The maximum mark for this examination paper is **65 marks**.

Section A

NOTES

1. **Figure 1: Soil composition triangle**

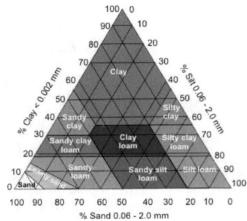

(Source: ADAS http://web.adas.co.uk/)

(a) (i) **State** which soil type is made up of 15% clay, 35% silt and 45% sand.

[1]

....

....

(ii) **Identify <u>two</u>** factors that determine the primary productivity of a soil.

[2]

....

....

Figure 2: The infiltration curves for sandy, loamy and clay soils

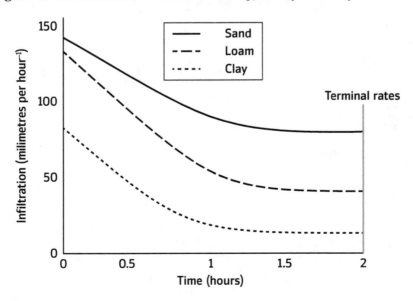

(b) (i) **Identify** the type of soil with the highest infiltration rate. [1]

....

....

(ii) Soil is a system in which transfers and transformations take place. Movement of water through the soil is an example of a transfer. **Identify two** transformation processes that take place in the soil. [2]

....

....

(c) **Outline** how deforestation can lead to soil degradation. [2]

....

....

....

....

2. **Figure 3: Concentrations of greenhouse gases from 1980 to 2015**

(Source: Adapted from www.esrl.noaa.gov/gmd/aggi/)

(a) (i) **Outline** the role that increased greenhouse gas emissions on global average temperature. [2]

....

....

....

(ii) **Calculate** the percentage increase in Annual Greenhouse Gas Index (Figure 3) between 1980 and 2010. [1]

....

....

....

....

NOTES

(b) Carbon dioxide, methane and nitrous oxide are considered as greenhouse gases.

(i) **State two sources of methane.** [2]

1. ....

....

2. ....

....

(ii) Apart from the gases mentioned above, **identify two** gases involved in the greenhouse effect. [2]

....

....

(c) **Outline two** ways in which carbon dioxide emissions can be reduced. [2]

1. ....

....

2. ....

....

3. **Figure 4: Stages of succession in a forest at times A, B, C and D**

A   B   C   D

(a) (i) **Identify** the type of succession shown in Figure 4. [1]

....

....

(ii) **State** the type of successional community shown at time D. [1]

....

....

(b) *K*-strategist species are often representative of climax latter stage succession.

Outline the characteristics of a *K*-strategist species. [3]

....

....

....

....

....

NOTES

(c) Biomass accumulation and productivity change in communities during succession.

Explain the differences in biomass accumulation and productivity in communities C and D. [3]

....

....

....

....

....

....

NOTES

Section B

Choose any **two** questions from a list of three. (Remember, in the real exam you will choose two from a list of four.)

4. (a) Conservation methods often involve habitat conservation, species conservation or sometimes a mixture of both.

   With reference to named examples, **distinguish** between 'species conservation' and 'habitat conservation'. **[4]**

   ....

   ....

   ....

   ....

   (b) **Suggest** a procedure that could be used to assess water quality using biological indicators. **[7]**

   ....

   ....

   ....

   ....

   ....

   ....

   ....

   ....

   ....

   ....

   ....

   ....

   (c) Around half of the tropical forests present at the beginning of the twentieth century had disappeared by the start of the 2010s, with the greatest loss during the 1980s and 1990s.

   Discuss possible reasons for the changes in the situation between the 1980s and 2010. **[9]**

   ....

   ....

   ....

   ....

   ....

   ....

   ....

   ....

NOTES

Answer lines continue on next page

....

....

....

....

....

....

....

....

....

....

5. (a) Substances that cause pollution are known as pollutants. Pollutants can either be primary or secondary.

Using named examples, **distinguish** between primary and secondary pollutants. **[4]**

....

....

....

....

(b) The amount of primary pollution released is usually at a maximum during rush hours (early morning and late afternoon). Despite this, photochemical smog is often greatest in the early afternoon. **Explain** why this is so. **[7]**

....

....

....

....

....

....

....

....

....

....

....

....

....

....

....

(c) Acid rain pollution can be managed at three levels: altering human activities causing the pollution, regulating and reducing at point of emission, and clean up and restoration.

With reference to either altering human activity or regulating and reducing emissions, **evaluate** the success of management strategies at either level. [9]

6. (a) Cork from cork trees was once the only method used by humans for stoppers in wine bottles. Modern alternatives made of plastics and rubber are now commonly used.

Using a named example, **outline** the dynamic nature of natural capital. [4]

(b) With reference to a named example, **explain** how the production process used in obtaining renewable natural capital could make its use unsustainable. [7]

Answer lines continue on next page

NOTES

(c) There are limits to the resources that humans can harvest from the planet.

Discuss the problem involved in the application of carrying capacity for human populations. **[9]**

RESOURCE BOOKLET

Set D: Paper 1

This Resource Booklet is for use with Set D, Paper 1 on page 88

Figure 1(a)
Map showing location of Venezuela

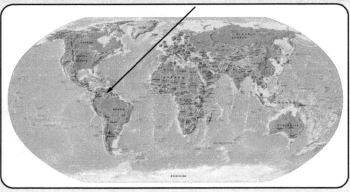

(Source: Central Intelligence Agency)

Figure 1(b)
Fact file on Venezuela

- Total area: 912,050 km².
- Land area: 882,050 km².
- Water area: 30,000 km².
- Irrigated land: 10,550 km².
- Climate zone: tropics.
- Venezuela's natural resources include petroleum, natural gas, iron ore, gold, hydropower and diamonds. It has the world's largest oil reserves. Its economy depends mainly on the petroleum sector and manufacturing.
- Petroleum products account for almost all of Venezuela's export earnings.
- Venezuela is prone to the following natural hazards: floods, mudslides and periodic droughts.
- Venezuela is one of the most urbanized countries in South America.
- The majority of Venezuelans live in the cities, especially Caracas, which is the largest city.
- In 2015, access to water supply and sanitation was 93%. Venezuela achieved the United Nations' Millenium Development Goals (MDGs) for water and sanitation ahead of time.
- According to 2015 estimates, 30.1% of the rural population in Venezuela had improved sanitation facility access compared to 2.5% of the urban population.

(Source: www.cia.gov/library/publications/resources/the-world-factbook/geos/ve.html)

Figure 1(c)
World map showing the location of Venezuela

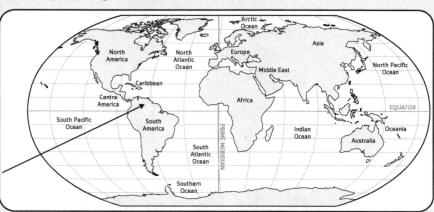

Figure 1(d)
Map of the World's Six Major Biomes

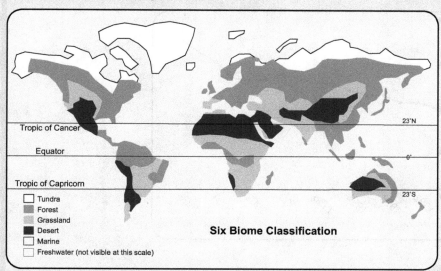

(Source: N Gardner, Four Corners Education, fourcornerseducation.net)

Figure 2(a)
Change in Venezuela's population over time

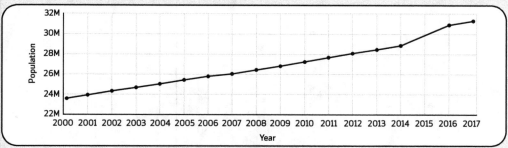

(Source: Adapted from www.cia.gov/library/publications/the-world-factbook/)

Figure 2(b)
Distribution of the population by age and sex, Venezuela, 1990 and 2015

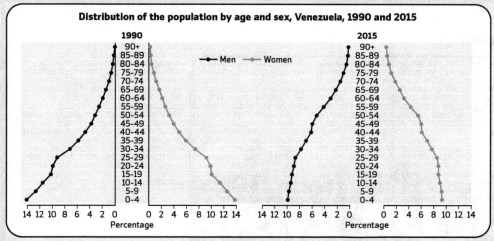

(Source: Adapted from UN Population and Statistics Divisions)

Figure 2(c)
Fact file on Venezuelan population demographics

- Venezuela has seen a decline in fertility and mortality over the past 25 years.
- Its annual population growth is 1.3%.
- Its birth rate is 19.0 per 1,000 population.

(Source: The World Bank: SP.POP.GROW and SP.DYN.CBRT.IN, https://data.worldbank.org/)

Figure 3(a)
Hydroelectric power consumption in Venezuela by year

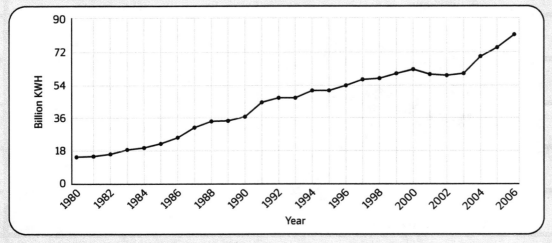

(Source: United States Energy Information Administration)

Figure 3(b)
Percentage of total energy use from renewable and nuclear energy from 1990-2015

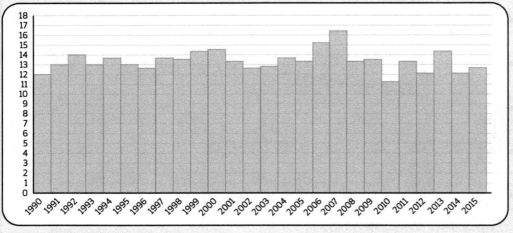

(Source: data from https://data.worldbank.org)

Figure 4(a)
Guri Dam on the Caroni River, Venezuela

(Source: Wikimedia Commons)

Figure 4(b)
Fact file on Guri Dam

- Guri hydroelectric power plant is situated 100 km upstream of the Caroni River.
- The plant provides 12,900 GW/h of energy, about 73% of Venezuela's energy needs.
- The government of Venezuela set up the plant to minimize the amount of energy produced from fossil fuels.
- The plant is the third-largest power plant in the world.
- Water levels in Guri Dam have fallen drastically. This could result in a power crisis in the country if the trend continues.

(Source: www.power-technology.com/projects/gurihydroelectric/)

Figure 5(a)
Grassy marshland with standing water

(Source: Olha Solodenko, Shutterstock)

Figure 5(b)
Swamp in Gran Sabana, Venezuela

(Source: klemen cerkovnik, Shutterstock)

Figure 5(c)
Fact file on the water situation in Venezuela

- Most rivers are in the south of the country while most inhabitants live in the north.
- Venezuelans have to depend on tanker trucks to supply water. They also have to deal with water rationing and contaminated water.
- Industrial and agricultural waste find their way into a reservoir that supplies 3 million people with water in major cities and towns in the country.
- Water crisis was brought about by, among other factors, changes in rainfall patterns caused by El Nino and La Nina phenomena, and mismanagement of available water.
- With the water crisis, health officials have warned of a risk of an increase in the mosquito population. This could increase the transmission of Zika, a mosquito-borne virus that causes brain damage in babies.
- Mosquitoes breed in stagnant water.

(Source: Adapted from www.ipsnews.net/2014/06/venezuelans-thirsty-in-a-land-of-abundant-water/)

Figure 5(d)
Zika virus infection pathway

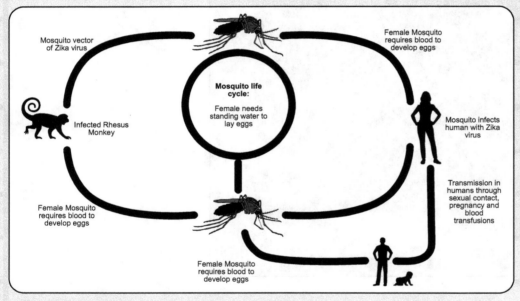

(Source: N Gardner, Four Corners Education, fourcornerseducation.net)

Figure 6(a)
Fact file about forests in Venezuela

- Venezuela lies just north of the equator between 0° and 16°N.
- Much of the country below 800 m experiences average annual temperature of around 27°C.
- Many areas experience over 2,000 mm of rainfall each year.
- Forests account for 51% of land use in Venezuela with a total of 466,869 km^2. The majority of these are rainforests.
- In 2012, forests covered close to two-thirds of the country's land mass. Of these, one third is found in Bolivar.
- Venezuela is one of the ten countries with the most biodiversity on Earth.
- Among the species found in the forests of Venezuela are:
    - 21,000 plant species
    - 353 mammal species
    - 323 reptile species
    - 1400 bird species
    - 288 amphibian species.

(Source: https://rainforests.mongabay.com/20venezuela.htm)

Figure 6(b)
Venezuela forest cover in 2012

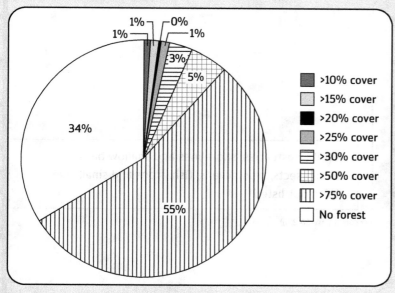

Legend:
- >10% cover
- >15% cover
- >20% cover
- >25% cover
- >30% cover
- >50% cover
- >75% cover
- No forest

Pie chart values: 1%, 1%, 0%, 1%, 3%, 5%, 34%, 55%

(Source: https://rainforests.mongabay.com/20venezuela.htm)

Figure 7
Photographs of examples of species found in forests in Venezuela:

(a) Giant otter (*Pteronura brasiliensis*)

(Source: Bernard DUPONT, Giant Otter (Pteronura brasiliensis) with a Vermiculated Sailfin Catfish (Pterygoplichthys disjunctivus), flickr.com (CC BY-SA 2.0))

Habitat: rivers, lakes, streams, creeks and springs
Diet: fish
Threats: killed by fishermen since they are competitors for fish
IUCN: endangered

(b) Black caiman (*Melanosuchus niger*)

(Source: David Stanley, Black Caiman, flickr.com (CC BY 2.0))

Habitat: slow-moving rivers, streams, lakes
Diet: fish
Threats: hunted illegally for meat and leather; their only other predators are Jaguars
IUCN: low risk

(c) Scarlet ibis (*Eudocimus ruber*)

Habitat: estuaries, shorelines and shallow bays
Diet: insects, crustaceans, fish, frogs and small snakes
IUCN: not listed

(Source: Fernando Flores, Scarlet Ibis | Corocoro Colorado (Eudocimus ruber), flickr.com (CC BY-SA 2.0))

(d) *Nectandra aurea*

- Endemic to Venezuela
- Used as timber
- Fruits are food for some bird species and also used as medicine by humans

(Source: Wikimedia Commons, Jorge Vallmitjana)

(e) Orchids (*Orchidaceae*)

- Grow in Venezuelan forests and lowland marshes
- Known as the national flower of Venezuela
- Valued for their colour and fragrance
- Believed to have medicinal properties

(Source: sunoochi, Cattleya mossiae fma. coerulea 'Blanca' C.Parker ex Hook., Bot. Mag. 65: t. 3669 (1838), flickr.com (CC BY 2.0))

Figure 8(a)
Trends in maternal mortality rates (MMR) 1990 to 2015

Country	1990	1995	2000	2005	2010	2015	% change in MMR (1990–2015)	Average annual % change
Cuba	58	55	43	41	44	39	-32.8	-7.1
Costa Rica	43	44	38	31	29	25	-41.9	-10.0
Argentina	72	63	60	58	58	52	-27.8	-6.2
Panama	102	94	82	87	101	94	-7.8	-1.1
Colombia	118	105	97	80	72	64	-45.8	-11.5
Saint Lucia	45	43	54	67	54	48	6.7	2.9
Brazil	104	84	66	67	65	44	-57.7	-14.9
Chile	57	41	31	27	26	22	-61.4	-16.9
Jamaica	79	81	89	92	93	89	12.7	2.5
Mexico	90	85	77	54	45	38	-57.8	-15.4
Peru	251	206	140	114	92	68	-72.9	-22.8
Ecuador	185	131	103	74	75	64	-65.4	-18.4
Venezuela	94	90	90	93	99	95	1.1	0.3

(Source: data from https://data.worldbank.org)

Figure 8(b)
Incidents of malaria in Latin America and the Caribbean

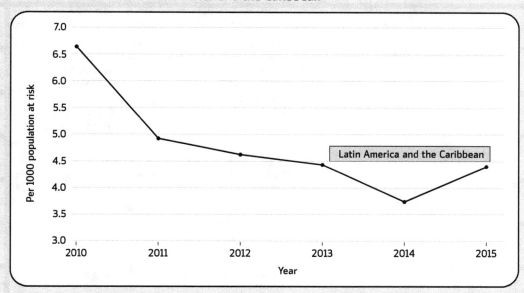

(Source: https://data.worldbank.org)

- In Venezuela, malaria cases have increased annually since 2008.
- In 2015, the country reported the highest number of cases in 50 years.
- The situation is made worse by mining activities and declining vector control interventions.

(Source: www.paho.org/hq/dmdocuments/2017/2017-cha-report-situation-malaria-amer-00-15.pdf)

Answers

SET A

Paper 1

1. • Tropical Evergreen Forest
   • Broadleaf Forest
   • Mangrove Swamp

   [Any two from the answers above, but you **must** use the exact words that are in the Resource Booklet. So: Tropical Evergreen Forest, **not** just Evergreen Forest. You get one mark for two correct answers but zero marks for only one correct.] **[1]**

2. Chiang Mai is in a bowl that is surrounded by mountains.

   [Text for this answer comes straight out of the Resource Booklet and, as a statement, needs no changing from what is in the booklet.] **[1]**

3. • Different tribes settle in different parts of the mountains, so their impact is spread over a wide area.
   • Tribes clear the forest for agriculture using slash-and-burn techniques. When the area becomes unproductive, they move to a new area where the forest is then lost.
   • An increase in population numbers means that the tribes need more space. As they produce their own food, they take more and more land to support the tribes. **[2]**

   [Only two answers are needed. Each answer must be clear and must have an identified factor and the reason why. Be careful you don't repeat the same answer just with different wording – you need to include different factors.]

4. $\dfrac{(1166978 - 953125)}{10}$ = 21385.3 people/year

   [Units **must** be included to gain the mark.] **[1]**

5. Any two of the following answers:
   • There are many endemic species in the area.
   • The area has high biodiversity.
   • There is a large threat from human activity in the area.
   • There is a large threat from population growth.
   • There is a large threat from land-clearance in the area.
   • There is a large threat from habitat destruction in the area.
   • The area provides economic benefits or natural capital tourism and food.
   • The area provides ecological services (with an example, such as being a carbon sink, providing flood control or the idea of forests as centres to produce oxygen).
   • The forest in the area has intrinsic value (this relates very strongly to Topic 1).

   [You can have any two of these answers, but you cannot have aesthetic/intrinsic value on its own unless it is referenced against, for example, tourism. This is the same for answering 'ethical reason' on its own – this is too wide and would need qualifying. Because the question is about biodiversity, just stating that the area contains endangered species is not enough on its own to get a mark.] **[2]**

6. One answer from the following:
   • There has been a reduction in population numbers.
   • The population size is small.
   • There are few mature individuals in the population.
   • The species has a restricted geographical range.
   • The places where the species is found are reducing in number.
   • There is habitat loss affecting the populations range.
   • There is a probability that without intervention the species could become extinct. **[1]**

7. Choose from Asian elephant or Indochinese tiger:

   Asian elephant
   • Protecting the Asian elephant promotes ecotourism.
   • It is an endangered species that is known to be endangered by most of the public.
   • It is a species that can be used successfully for publicity and fund raising.
   • Elephants are aesthetically attractive and people like elephants.

   Indochinese tiger:
   • Protecting the Indochinese tiger is a potential attraction for ecotourism.
   • Because it is a predator, its loss could largely negatively impact other species in the ecosystem.
   • The Indochinese tiger is found only over a small range so is very representative of the Indochina Highlands.
   • It is a species that would be useful for publicity and fundraising as people find tigers attractive and they like them.

   [As this question is about a flagship species, it needs to combine with the idea that by protecting a flagship species you will also protect other species within the same habitat as the flagship species. If you answer about a plant species, you get 0 marks as plant species are not regarded as flagship.] **[3]**

8. • The Great Hornbill males have a home-range that reduces from nearly 15 km to nearly 4 km during their mating season. Fragmentation could mean there are fewer suitable areas that are big enough to support the males in the mating season.
   • With habitat fragmentation, the area of the undisturbed forest becomes more and more broken up. This means that the distance between suitable sites may increase and be too far to travel between.
   • This potentially leads to males being unable to access females if the forest fragments are more than 4km apart. This may result in a reduced population. **[4]**

9. (a) • After the rice has grown and been harvested, there is rice straw left on the land.
   • Burning is a fast, effective and convenient way of removing the rice straw.
   • Burning the straw returns nutrients back to the soil – this helps with improving the soil.
   • Burning the straw reduces the need for additional fertilizers so saves on the production costs.
   • Once the straw has been burnt, the soil is prepared, including the ash, and then the next crop is planted without the need for heavy ploughing or machinery **[2]**

   (b) • Despite the fact that burning the straw puts some nutrients back into the system, a lot of carbon is lost as carbon dioxide or carbon monoxide from the system.
   • Burning the straw creates pollution in the form of smoke particles.
   • The fires to burn the rice straw could get out of control and destroy nearby habitats. **[2]**

10. • A lack of air movement creates a layer of dense, cool air, which is trapped beneath a layer of less dense, warm air.
   • This causes a concentration of air pollutants to build up near the ground, rather than being moved away by normal air movements. **[2]**

11. • The general trend was annual fluctuations, with a spike event where readings reached over 100 μg/m³.
   • Before 2011, peaks reaching over 200 μg/m³ were rare. But, in 2012, the peak rose to nearly 500 μg/m³. **[1]**

12. (a) • An increase in the numbers of vehicles in the city has increased the amount of air pollution produced at street level.

  • Population growth means more people need more food, which requires more land and results in the burning of more rice straw.

  • Less tourism because of pollution means more people depend on the rice harvest for economic security, so the amount of rice grown increases and more straw is burnt. **[2]**

 (b) • Reductions in pollution on a local level are being addressed by burning bans, so these require effective enforcement.

  • Baling and selling rice straw can make money for farmers, but it requires machinery. This costs more and produces greenhouse gases itself.

  • The creation of Biochar still needs machinery to remove rice straw, but it can be a replacement for fertilizers and does not produce greenhouse gases.

  • Reducing car usage in the city centre would help with ozone build-up. This can be achieved by improving public transport, encouraging cycling, etc. **[2]**

13. • With ecotourism there are increased revenues to invest back into conservation.

  • Ecotourism raises awareness of conservation issues. This leads to greater support and public engagement with wildlife conservation.

  • Ecotourism increases a consideration of wildlife as an asset that requires looking after.

  • If the local population have jobs in the ecotourism industry, they are less likely to engage in unsustainable rice farming. This will reduce the demand for land. **[1]**

14. • A growth in the number of tourist sites (or hotels or homestays) could cause a loss of habitats in the forest.

  • Increased ecotourism could lead to the creation of roads. This could lead to habitat fragmentation.

  • Noise pollution from tourism near wildlife areas can disrupt mating.

  • Coming into close contact with tourists can alter animal behaviour.

  • Litter left by tourists can degrade the environment and harm wildlife.

  • Increased tourism puts greater demand on limited freshwater that is unsustainable.

  • As more people gain greater access to wildlife areas, increased poaching could occur. It is easier for poachers or plant collectors to get into remote areas with new infrastructure.

  • Ecotourism could increase demand for services in the area that cause deforestation or the use of unsustainable resources (e.g., illegal cutting of the forest for fuel).

  • Wildlife near tourist attractions may be treated inappropriately or become dependent on people for food.

  [Any two answers here are required] **[2]**

15. *[In your answer you will get 5 marks for argument and 1 mark for conclusion.]*

Arguments FOR development away from traditional rice farming [4 max]:

• Reducing the amount of rice farming that takes place can lead to a reduction in:

  o Deforestation

  o Pollution from straw burning

  o Soil erosion

• Increases in tourism/ecotourism could increase investment in conservation. This could help limit deforestation and help to protect wildlife.

• Could lead to increased environmental awareness.

• Could lead to alternative farming practices that become attractive ecotourism destinations.

• Increase in jobs associated with tourism could reduce the dependence on rice farming.

• Less dependence on rice farming could lead to increased education as less labour is needed on the fields.

• This could lead to lower population rises with better family planning. Less population growth could lead to a lower footprint for the people in the area.

• More people may stay in the area with better paid jobs in tourism than move to cities.

Arguments AGAINST development away from traditional rice farming [4 max]:

• Changing alternative rice farming methods, such as more intensive chemical, fertilizer-based farming, could lead to soil degradation.

• Eutrophication/habitat damage.

• Increased tourism could lead to greater deforestation as land is used for hotels, shops, homestays and resorts.

• Growth of tourism could lead to increased demand on water supplies and lead to droughts.

• This could lead to increased problems of sewage if development occurs faster than infrastructure development.

• Damage from building access roads could lead to loss of forest.

• Increased number of tourists could damage the ecosystem through trampling, etc.

• Increased number of tourists could mean more pollution, such as solid waste, plastics from food outlets, etc.

• Change in lifestyle of local people is likely to increase consumerism.

• Food production may become more intensive.

• More food may need to be imported into the area from outside, increasing the footprint of the area.

[You do not need to include all of the possibilities for and against, but you must have at least 5 points in total from both for and against to be awarded 5 marks.]

Conclusion example [1 max]

• Changing away from rice farming towards ecotourism should lead to greater sustainability. However, there needs to be caution as ecotourism could also bring its own problems.

[The conclusion needs to be valid and balance between both for and against arguments. It needs to have explicit details that give evidence to the conclusion.] **[6]**

Paper 2: Section A

1. (a) Temporal **[1]**

 (b) • There are a greater number of species/habitats/niches/genetic diversity in climax communities than in pioneer communities.

  • The gross productivity or stored biomass is higher in climax communities than in pioneer communities.

  • There are more complex energy pathways or food webs found in climax communities than in pioneer communities.

 [You need two answers with a reason.] **[2]**

 (c) As a community moves through the successional stages from a pioneer community to a climax community, primary productivity will increase as the stability of the community increases, allowing the producers to increase in type, abundance and size. Or this can be in relation to increased complexity of the habitat as it gets older. **[2]**

 (d) • K-strategist species tend to produce a small number of offspring, which increases their survival rate and enables them to survive in long-term climax communities.

  • K-strategists produce fewer offspring so have slow growing or stable populations.

  • K-strategists often have narrower distributions that are linked to stable environments.

- K strategists mature slowly, so can exploit resources over a long period of time. Therefore, they need more stable conditions.

- K-strategists often have narrow niches that may be fulfilled by a particular climax community. **[2]**

(e) • Altitude

- Latitude

- Tidal level

- Soil nutrients/soil pH **[1]**

2. (a) Renewable natural capital is natural capital that can be used again (it is not used up). It can be used sustainably but it can also be used unsustainably where its availability will be reduced but not totally depleted. **[1]**

(b) • Luxembourg

- Croatia

- Lithuania **[2]**

(c) • Bulgaria

- Estonia

- Finland

- Czech Republic **[2]**

(d) Cypress as it is almost totally dependent on oil. **[1]**

(e) (i) Technocentric as this is using technology to take a waste product and make further use of it. **[1]**

(ii) There are a number of major environmental issues relating to this generation of energy, as it still produces a significant amount of air pollutants when incineration takes place. **[1]**

3. (a) Sunlight / solar energy / solar insolation / the sun **[1]**

(b) • The build-up of toxins in an individual over its lifetime is Bioaccumulation.

- The build-up of toxins as they move through the food chain is Biomagnification. **[1]**

(c) When an aerosol DDT gets moved by the wind, if DDT particles get into the upper atmosphere they can be transported from the source to be deposited at a different point such as the Arctic. **[2]**

(d) • For example, the fish is eaten by the crayfish, but the crayfish will eat more than one fish.

- For every small fish that is eaten, the crayfish takes on one unit of mercury.

- As the crawfish grow and age, the concentration of DDT in their bodies will increase.

- The polar bears eat a lot of fish (so they take in the DTT from more than one fish) meaning the concentration is magnified.

- Polar bears are large and long lived, so the DTT will accumulate over time and the concentration will increase as the polar bears eat more fish.

- This means the polar bears, which are in tropic level 4 or 5, will have the highest concentration of DDT in the food chain. **[5]**

Paper 2: Section B

4. (a) • r-strategist species produce large numbers of offspring.

- They colonize new habitats quickly, making use of scarce or limited resources.

- K-strategist species produce a small number of offspring.

- They require more stable conditions to thrive.

[You need to mention both the number of offspring and the use of resources to get the full marks.] **[4]**

(b) • Succession is the change in a community of organisms over time.

- This starts from newly developed or disturbed habitats that slowly become more stable.

- r-strategists are most abundant in the pioneer stage because of their quick reproduction and rapid growth and development.

- This allows the species to recover quickly from disturbances.

- They are also good at adapting and can potentially deal with the level of change that happens in the early development stages.

- K-strategists take a long time to mature and breed.

- They are, therefore, not adaptable to change as the habitat moves through stages of succession, becoming more complex and more stable. **[7]**

(c) Early stages of succession tend to be marked by few species within the community. As the community passes through subsequent stages, the number of species in the community increases. Very few pioneer species are ever totally replaced as succession continues but the number of species increases as more species enter the community. The result is increasing diversity (for example, more species). This increase tends to continue until a balance is reached between possibilities for new species to establish, existing species to expand their range and local extinction.

Diversity within systems leads to stability and greater resilience to tipping points.

As the community becomes more complex, the greater range of nutrient and energy pathways develop. This leads to greater stability. Greater connections are created in food webs so that the loss of one species can be mitigated by other species in the community. So, the stability of a community is connected directly to its stage of succession – moving from less stable pioneer communities to more stable climax communities.

Human activity can alter the course of succession, either by stopping the successional process or by destroying the climax. This can either create cyclic conditions if the disturbance is repeated at time intervals, such as in slash-and-burn farming where the land is cleared by fire, planted for a few years and then abandoned to start succession again. Or a new climax state can be maintained by continued human activity, such as clearing forests for grassing land for cattle. Continually grazing cattle will stop the regrowth of forest, and it could also lead to other effects such as soil erosion that reduce the stability of the ecosystem. **[9]**

5. (a)

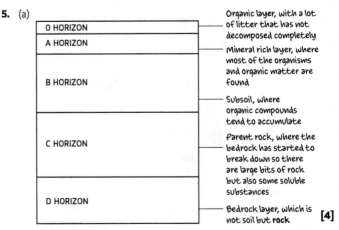

O HORIZON	Organic layer, with a lot of litter that has not decomposed completely
A HORIZON	Mineral rich layer, where most of the organisms and organic matter are found
B HORIZON	Subsoil, where organic compounds tend to accumulate
C HORIZON	Parent rock, where the bedrock has started to break down so there are large bits of rock but also some soluble substances
D HORIZON	Bedrock layer, which is not soil but rock

[4]

(b) Leaching is the downward movement of water through the soil that is also dissolving minerals and removing them from the soil. This can take place in the B horizon, as shown in the diagram. Leaching increases with an increase in the permeability of soil. Sandy soil will leach the most nutrients, whereas a more clay soil will leach fewer.

However, if the clay content is too high, the minerals become bound to the clay surface and are therefore not available to use. Locations with monsoon seasons experience high leaching on a regular basis. Sloping surfaces will also result in more minerals being leached from the top to the bottom of the slope. This can be limited using water conservation methods such as terracing or contour ploughing in steep places. If vegetation is present then leaching can be reduced as plants absorb nutrients, but if there is no vegetation present then leaching can become considerably worse.

[Or any other correct answer.] **[7]**

(c) [Depending on the commercial farming system studied, your answer could include the following:]

Soil conditioners (organic materials and lime)

- Conditioners such as lime improve the quality of soils and reduce pH, which helps increase yields.

- Adding organic matter and composting adds nutrients back to the soil that are lost with the removal of crops and biomass in harvest and grazing. Addition of organic matter replicates the nutrient cycles. Adding organic matter also helps with water retention in the soil.

Wind reduction techniques

- Provision of shelter through growing hedges, etc. provides shade and helps slows down the drying out of the soil and reduces the effect of the wind carrying dry soil particles away.

Cultivation techniques (terracing, contour ploughing, strip cultivation)

- Terracing reduces both soil erosion and water run-off from the land.

- Contour ploughing produces mini terraces with the same effect.

[In your evaluation, you need to consider how successful any of these techniques are in the situation you studied. For example, terracing in tropical rice growing areas has been very successful at extending the growing area up very steep hill sides.

You need to mention any problems they may create – such as overuse of manure as additional organic material can lead to eutrophication.

You could also consider if these methods are sustainable.]   **[9]**

6. (a) • Economic growth throughout the population gives greater economic security and independence, reducing the need for large families as insurance policies.

- Providing education for women leads to greater independence – both reproductive independence and economic. With economic independence, women can choose to have children later or smaller families.

- Improving agricultural production systems increases the amount of food available, reducing the need for large families as labour.

- Improving health services and sanitation can reduce the death rate and lead to population increase.   **[4]**

(b) In many countries, one of the main changes to birth rates has been the improved status of women in society, resulting in women receiving better education and healthcare. Often, a better education results in more women choosing to work rather than have large families. The increase in healthcare has allowed women to have safer pregnancies and infants a better chance of survival.

In subsistence economies, children are valuable labour. But a reduction in infant mortality has resulted in more children surviving for longer. This improvement reduces the need to have so many children; this is seen in many African countries such as The Gambia or Gabon. In developed countries, the birth rate is dropping due to the cost of having children rising greatly, for example, in Japan and the United States.   **[7]**

(c) [You answer will depend on the local area you have studied.]

'Carrying capacity' is the maximum number of species and individuals a certain area can sustainably maintain. Some parts of the world are populated above their carrying capacity, whereas others are not.

Populations are growing at different rates in different countries.

In developing countries, the populations are fast growing due to increased resources, but then have increased needs of education, employment, contraception and knowledge. There are more females in the workforce, which leads to a potential reduction in childbirth. Many countries live beyond the resources available, which results in unsustainable practices.

The rate of expansion is different in developing and developed countries. In developing countries, expansion and development

is happening at a more accelerated rate as they very quickly shift from limited development to Westernized development and the proportion of people in these areas further enhances the problems. Self-regulation of resource use is difficult to achieve.

Technology has allowed human populations to live beyond their carrying capacity. The ability of humans to switch resources in times of scarcity has also allowed this to happen. This gives access to a much greater range of resources. Humans are also able to move resources around, meaning plentiful resources in one place can be moved to where they are scarce. Food export and import is an example of this. Technological changes also change what resources are available.

However, over-consumption has led to unsustainable practices. Pollution, climate change, etc., are a result of population growth and over consumption.

[Your answer needs to discuss how successfully your community has applied this.]   **[9]**

SET B

Paper 1

1. Conifer Forest

   Broadleaf Forest

   Grasslands

   Shrub vegetation

   [One mark for any two correct answers but give yourself a zero mark if only one answer is given.]   **[1]**

2. • Birth rate is higher than the death rate according to the data, with most of the population under 40.

   - There are more males than females in the population aged between 25 and 40.

   - The birth rate is has slowed over the last 30 years, which will result in a larger ageing population over time.

   - The population is slowly moving from rural to urban areas.

   [You need to have identified **two** trends in the population data.]   **[2]**

3. • Bhutan has a declining birth rate.

   - It has low death rate.

   - More people living to old age.

   [You need to have any **two** to be awarded two marks.]   **[2]**

4. rNIR = (Crude Birth Rate – Crude Death rate)/10 = (17 – 6.4)/10 = 1.06 %

   [Give yourself zero marks if you forgot to include the %.]   **[1]**

5. • Bhutan needs to create a more balanced approach to renewable energy production to make sure it is not overly reliant on hydroelectricity.

   - Increase the use of hydropower as only around 5% of the country's resources are currently being used.

   - The development of solar power could be a viable option for Bhutan.

   - More wind power generation stations to take advantage of the ideal mountainous location.

   - Bhutan could reduce the export of electricity to India as currently 70% of Bhutan's electricity goes to India.

   - Explore the possibility of using biogas.

   [You need to have two suggestions that would improve Bhutan's independence for both marks.]   **[2]**

6. Bhutan's ecological footprint has remained relatively stable during the period, though there was a rise in EF both between 1990 and 1996 and again after 2010.   **[1]**

7. The availability of cable TV and cellular telephone services after 2012 will have increased the footprint, both before, as the infrastructure is put into place, and afterwards as availability will have led to increased consumerism.   **[1]**

   [Or words to that effect.]

8. • Bhutan has a strong link to the environment through its Gross National Happiness and the fact that it is carbon negative.

• Buddhism is strongly linked to the environment and as Bhutan's main religion, Buddhism supports the environmental vision of the country. **[2]**

9. • Bhutan is a country that has prioritized maintaining over 60% of its land as forest.

• This has provided unique areas that have had little disturbance and very little input from human society.

• Large mammals need large home ranges; these are provided in Bhutan due to the limited habitat fragmentation.

• 50% of the country is environmentally protected so hunting and poaching are limited.

• Bhutan is a small, isolated country in the Himalayas, so access is limited which helps to prevent disturbance of the ecosystems and helps prevent poaching.

• The lack of disturbance in Bhutan's habitats means there are robust food webs that function well enough to support numerous adult individuals at the top of the food chain.

[You need to explain the link between high numbers of large protected species and the measure that allow the protection.] **[5]**

10. • Building dams requires the removal of trees within the immediate area. This can lead to landslides due to the lack of trees anchoring in the soil.

• Damming rivers can result in upstream flooding and downstream losses of water.

• Building dams may require the building of access roads, this can create habitat fragmentation. **[1]**

11. • Building dams requires the removal of trees within the immediate area so you get habitat destruction.

• Building dams could lead to a loss of nesting sites for birds and a loss of bird species, because of a loss of trees.

• The lakes behind dams flood the existing habitats so they are lost which results in a loss of diversity.

• Possible reduction in fish species because dams act as a barrier to season migration.

• Fish species that require flowing water could disappear.

• Removal of trees during construction can lead to landslides due to the lack of trees anchoring in the soil, this could lead to eutrophication in the water and a loss of species.

• Damming rivers can result in upstream flooding and downstream losses of water, which could alter the fish populations.

• Fish migration patterns will be disturbed, which could affect the Asiatic black bear populations.

• Access roads can create habitat fragmentation, smaller habitats for species that depend on the forest.

• Access roads could lead to increased poaching and loss of species.

*[Award yourself any **three** answers or any sensible alternative answers that explain rather than just describe.]* **[3]**

12. An EIA helps in the decision-making process for a project by providing information about possible consequences and impacts on the environment, society and economy that may occur because of the project. **[1]**

13. Before the construction of the dam, in selected areas, use a random sampling technique with a quadrat to measure the abundance of species within the area. Take enough samples to make sure a full coverage of all species is recorded. Calculate the mean (average) number of species per unit area from results. Using a Biodiversity Index (Simpson Index of Biodiversity) calculate the diversity index for the area. This provides the baseline data/index to compare all other later results. Following the construction of the dam, repeat the sampling surveys at 2-year intervals.

[As this is a ten year period, 2 or 5 year intervals would be sensible.] **[3]**

14. Increasing economic development could lead to the following.

• Increase in electricity supply to rural areas.

• Decreasing export of energy to India reducing the need for as many hydroelectric power stations.

• Reducing the need to subsidize electricity in rural areas allowing money to be used for other projects.

• Moving to other forms of electricity generation could lead to a reduction in the need for hydroelectric power.

• Possibly lifting of tourism restrictions could lead to increase the potential of the tourism industry. This could lead to increased importance on ecotourism. **[2]**

15. Increasing economic development could lead to the following.

• Any development that took place would result in a reduction of the pristine forest habitat that was present.

• Changes to economic development could be required energy production to increase development to keep up with demand for energy.

• This could reduce the level of energy exported to India reducing the countries income.

• This could increase the need to drastically increase supply through either more hydroelectric stations, wind turbines or even the use of fossil fuels.

• Possibly lifting of tourism restrictions could lead to increase the potential of the tourism industry. This could lead to a decline in habitats, and possibly a decline in religion and traditions in the country. **[2]**

16. For something to be sustainable, the amount or value of the resource being managed needs to be maintained, not reduced. Bhutan has a strict tourism policy that limits the number of visitors. This helps to eliminate effects such as habitat destruction because of overuse of trails, etc. Set daily fees cover accommodation, travel, food and a registered guide. This means that a person needs to be part of a designated tour in order to be a tourist in Bhutan. Small groups (fewer than three people) are charged additional money.

These measures have kept tourism low and controlled, as Bhutan is not accessible to everyone due to the cost.

While it is a good idea to limit the impact on the environment, this policy will also deter some visitors from coming to the country for a holiday and, therefore, possibly limit the amount of economic advantage of sustainable tourism that would allow local communities to move away from less sustainable agriculture and fishing.

With the locations and hotels limited, it is hard for the younger local population to develop a livelihood through tourism, and this could drive them to seek opportunities elsewhere. **[6]**

SET B: Paper 2

Section A

1. (a) 2002 **[1]**

(b) 700 – 100 = 600% increase in consumption. *[If you forgot to add the % then give yourself 0 marks for this question.]* **[1]**

(c) The number of households has increased significantly. The efficiency of each cooling system has not really improved during this time so the total amount of energy used continues to rise. **[1]**

(d) Ozone depleting substances (including halogenated organic gases such as chlorofluorocarbons – CFCs) are used in aerosols, gas-blown plastics, pesticides, flame retardants and refrigerants. Halogen atoms (such as chlorine) from these pollutants increase destruction of ozone in a repetitive cycle. **[1]**

(e) • Damages human living tissues in the form of sunburn and aging of the tissue lack of elasticity.

• Increasing the incidence of cataracts which if untreated can lead to blindness.

• Mutation during cell division which can cause cancers, skin cancer and other subsequent effects on health from being a cancer patient. **[2]**

(f) An illegal market for ozone-depleting substances persists and requires consistent monitoring. **[1]**

2. (a) Renewable natural capital can be generated and/or replaced as fast as it is being used. **[1]**

(b) Economic water scarcity is where water is present within the area but the population does not have sufficient infrastructure to be able to access it, e.g. Yemen, Libya, Western Sahara.

Physical water scarcity is where there is not enough water or no source of water, e.g. California, northern Africa, Morocco, UAE. **[2]**

(c) For example: The Red River in Hanoi, Vietnam (but can be any system you have studied).

- Population growth has meant that water is extracted from the higher reaches of the Red River as drinking water – high population growth results in more and more extraction. The water is also used by farmers to irrigate crops both around and in Hanoi itself.

- Rapid industrialization has also meant that the Red River is a major source of water for industry.

- The river has also become a sink for pollution as road run off effluent and sewage enter the river at different points this has led to eutrophication or toxicity increases in the water.

[This is a three mark question so you would be expected to have more than a couple of outlines] **[3]**

(d) • Building more reservoirs.

- Using desalination of sea water near the coast.

- Artificial recharge of aquifers by bringing water from areas without scarcity.

- Rainwater harvesting schemes, such as rainwater capture on the roofs of new buildings.

[Any two] **[2]**

(e) Inorganic fertilizer → Nitrate enrichment → Eutrophication

[1 mark for the parts being correctly labelled and 1 mark for the arrows being in the right direction] **[2]**

3. (a) Indicator species are species that are susceptible to changes in the environment and require a specific set of biotic or abiotic conditions to flourish. Changes in their population can be used to indicate changes in the environment. **[1]**

(b) • They are pollinators and without them many plants would not reproduce. **[1]**

- They provide honey as a food source for other organisms

- They are the prey for other trophic levels.

[Any one answer.]

(c) • Reduction in the number of plants that are pollinated, resulting in low crop yields.

- Loss of plant biodiversity, resulting in a knock on loss of herbivores.

- Potential increase in the populations of other pollinators as the bees disappear. **[2]**

(d) • Import bees from areas where there has been limited loss.

- Breed bees in captivity and release into the environment.

- Hand pollination by farmers using something like a paintbrush. This is very time consuming and would only work on a very small scale.

- GM crops to self-pollinate, although this will take time to develop to become reliable.

- Allow non-crop plants to grow around the crop to attract pollinators other than bees. **[4]**

Paper 2: Section B

4. (a) • Storages could include plants, animals, fossil fuels (*you will need to list the separate fossil fuels to get the full marks here*), in CO_2 in the atmosphere, or in limestone.

- Processes could include respiration, combustion, sequestration, fixation, volcanism, weathering, photosynthesis or decomposition. **[4]**

(b) The carbon cycle is key to the functioning of many natural processes, but due to its importance in human life our activities are changing the way this natural system functions.

Extraction and use of fossil fuels. The extraction of oil, coal and gas depletes the stored carbon. When these fossil fuels are burned they release CO_2 and the carbon is released into the atmosphere.

Trees and other plants release CO_2 as a waste product from respiration but they also remove carbon through uptake and sequestration during photosynthesis. Deforestation is reducing the level of sequestration taking place and therefore the amount of carbon being removed from the atmosphere.

[Or any other reasonable answer.] **[7]**

(c) Renewable energy is naturally generated, cleaner and more sustainable than using fossil fuels to create energy. It has a much lower impact on the natural carbon cycle.

Bhutan, which is in a mountainous region, has been able to develop a hydroelectric system. Bhutan uses little to no fossil fuel generated electricity.

However, landslides in the mountains are now more common due to tree removal for dam construction.

This can lead to the release of both carbon dioxide and methane from the soil which increase the amount of greenhouse gases in the atmosphere. As well as the movement of minerals including carbon through transport in river systems. It also reduces the store of carbon held in forest biomass as trees are removed.

Much of Bhutan only has limited access to electricity and, in some areas, the main fuel used is still wood or even animal dung. This produces carbon dioxide when burned.

[Or any other similar case study]

The use of renewable energies will help reduce the removal of carbon from the environment and reduce the increase in carbon dioxide in the atmosphere. The use of renewable energy makes Bhutan the most carbon negative country in the world. Overall this helps limit the severity of some aspects of climate change. **[9]**

5. (a) **In-situ:** The conservation of species in their natural habitats, considered the most appropriate way of conserving biodiversity.

Conserving the areas where populations of species exist not only helps conserve a particular species but also all the species in that ecosystem.

In-situ conservation may include the establishing of reserves and protected areas as well as corridors between protected areas.

Ex-situ: The conservation of species outside their natural habitats. This involves conservation of genetic resources, as well species.

Methods can include:

- Seed banks, sperm and ova banks – genetic stores

- In vitro plant tissue and microbial culture collections

- Captive breeding of animals

- Artificial propagation of plants (with the aim possible reintroduction into the wild)

- Collecting living organisms for zoos, aquaria, and botanic gardens for research and public awareness

Ex-situ conservation provides an "insurance policy" against extinction. **[4]**

(b) The garigue grasslands in the Mediterranean are disappearing because of rural depopulation. Schemes are in place to maintain the plant biodiversity by annually cutting back encroaching trees in the ecosystem.

To compare the biodiversity of an area being protected and one not protected, you need to survey both areas. You would need to identify all the plant species found in the area and sample them using a quadrat sampling method in a random grid making sure enough quadrats are placed to effectively sample the entire area. Abundance of each species would be calculated as number of individuals of each species per m^2. The results could then be compared using Simpson's Diversity Index, to show any difference in the diversity of the two areas. **[7]**

(c) Things your answer should consider depending upon the case study you use:

- Biodiversity is the number and relative abundance of living organisms in a given area.

- Biodiversity is negatively influenced by most human activities.

- Major causes of biodiversity loss include human population growth, deforestation, habitat fragmentation, agriculture, urbanization, trophy hunting, hunting and collecting for medicinal value.

- For example: Reforestation can help to reduce loss of forest and habitat fragmentation. This ensures that there is a future source of wood and other benefits related to increased biodiversity. However, planting young trees does not provide immediate replacement of the oxygen lost by removing mature trees. Also, the balance of the ecosystem will change due to disturbance in the habitat.

- Forests in Sweden have been harvested for their old growth wood in the past and many were replanted as plantation in the last 50 years. Some, however, were designated as nature reserves alongside remaining areas of old growth forest. Wildlife in these areas is protected but the complexity of the original forests is lost following replanting and fewer species are found in them than areas that have not been cut. However, where areas have been protected from the removal of old growth forest a wider range of species exist and more nutrient pathways are available. This creates a more complex environment that is more resilient to change and a more complex food webs exist.

- In the new growth areas, fewer higher trophic level species are found than in the old growth areas, and in the early stages of growth, reindeer and moose numbers need controlling as they can damage the young trees. Where old growth forest has been conserved the impact of man has been negligible. While the new growth reserves have a greater range of wildlife than the old growth, they are not as successful at retaining or regaining the original biodiversity, so have not been as successful at reducing human impacts as retaining the old growth forests in the area.

[You will need a similar EVALUATION. How successful has it been? What has made it successful? What has not made it successful?]
[9]

6. (a) Reducing the amount of consumption of new products; repairing and re-using items; making used items into something else that is useful (upcycling); recycling all plastic, tin and cardboard; and buying sustainable, reusable products, e.g. bamboo straws and glass jars. **[4]**

(b) Recycling is where certain items such as glass, plastics and paper are collected once they have been used and turned back into new items.

While this has been a way of dealing with some of our solid domestic waste for a long time, there is a limited number of different substances that can be recycled. Mixed material products and lots of types of plastic cannot be recycled, resulting in lots of waste that recycling will not reduce.

Recycling programs can be encouraged by educating people about first considering buying products that are sustainable. If those are not available, also by educating people about recycling the products that can be recycled.

Making it easier for people to recycle by having roadside collections or supplying waste bins specifically for material that can be recycled. Or collection points at certain places such as electrical stores for old batteries.

As a last resort, legislation can be used to force people to recycle. In some countries you can be fined if you put recyclable waste into the non-recycling collection. **[7]**

(c) The three levels of pollution management look at activities related to reducing the demand for products, reducing the release of solid domestic waste and cleaning up waste that has been created.

Reducing demand for products

Improve the quality and number of reusable items such as cutlery, bags, etc., for example, bamboo is extensively used in Asia for straws, cutlery, construction, and so on.

Reducing the release into the environment

Better collection and disposal of waste, more upcycling of items, e.g. plastic bottles are being used to make rafts and boats in some developing countries. Precious plastic is a global project showing people how to collect plastics and make floor tiles and furniture from them.

Clean up

The Great Pacific Garbage Patch is an example of a major impact. It is being cleared slowly by the collection of large waste from the surface. Beach clean-ups, etc., are smaller-scale examples.

[Or any other reasonable examples.]

In conclusion, it is clear that the ideal situation is that this issue is tackled first through reducing demand as successes in this area will automatically lead to reductions in both of the other management levels. **[9]**

SET C

Paper 1

1.
- Newfoundland is an island so only has a limited area. This means there may be fewer niches for species to exploit, so you end up with fewer species.

- As an island, species may need to migrate from the nearest continent, so only those that can successfully migrate can get there.

- Being very far north, species that live on Newfoundland need to be adapted to the possibly very cold winters and short summers, so the abiotic conditions limit the number of species.

[Any one answer] **[1]**

2. Competition

Predation

Herbivory

Mutualism

Parasitism

[Any two] **[2]**

3. Fishing

Mining

Agriculture/farming/crop farming

Tourism

[Any two] **[2]**

4. 108,860(km^2) × (60/100) = 65,316 km^2

Or

108,860(km^2) × 0.6 = 65,316 km^2

[If you forgot the units, give yourself zero marks.] **[1]**

5. Answer should include four trophic levels, such as:

White Spruce → Red Squirrel → American Marten → Red Fox **[2]**

6.
- The main income on the island comes from sea fishing, so the population needs to be near the coast to take part in fishing.

- Soil generally is poor so cannot be relied upon for income from agriculture, so there is no incentive to live away from the coast.

- The sea will help keep the climate of the coast milder than inland so it may be easier to survive near the coast.

- As an island, transport links to the rest of Canada are easier near the coast, so people will settle where the transport links are. **[2]**

7. (a) **Advantages:**
- Factory fishing uses modern technology – super trawlers.

- These catch more fish.

- Allows fishing over a wider area. **[2]**

(b) **Disadvantages:**

- Resulted in overfishing.
- Unsustainable.
- No time allowed for fish populations to recover. **[2]**

8. (a) 2000–2007: Population <u>decreased</u> from 527,966 to 509,039;
2007–2016: Population <u>increased</u> from 509,039 to 530,305;
A <u>slight increase</u> between 2016–2017. **[2]**

(b) Population increase due to:

- Immigration of people to work in the area because of a return of fishing stocks.
- Greater income from reestablishment of fishing to support a larger family.
- There could have been incentives from government to have more children.
- There may have been higher birth rates than death rates.

[Any two sensible suggestions, they could be directly from the resources or from critically thinking about what could have made these changes.] **[2]**

(c) • Humans use a variety of resources so they can substitute one resource for another this means they can alter their niche requirements.

- Lifestyle affects resource requirement and use, so life changes can alter what people need.
- As technological developments change, so do the resources needed. An example would be increased agricultural production because of chemical fertilizers.
- Can import resources from where they are plentiful to where they are scarce. **[2]**

9 **RNC:**

Fishing

Agriculture

Forestry

NRNC:

Iron Ore

Gas

Oil **[1]**

10. The percentage of GDP that comes from agriculture is very low at only 1%.

Because of being near the Arctic, Newfoundland experiences long cold winters and short summers with a small growing season, which affects agricultural productivity. The soil is also not suitable for farming/agriculture. **[3]**

11. • Loss of habitat/biodiversity because of removal of existing community.

- Decrease in the *Braya* population because of:
  o noise pollution from heavy machinery
  o air pollution from dust created during mining.
- Loss of aesthetic value because of the mining.

[Any two for both marks.] **[2]**

12. • Movement of tectonic plates produce barriers such as islands, mountain ranges and valleys.

- These lead to isolation of gene pools.
- This prevents gene exchange between isolated populations.
- Over a long period of time – as the gene pool of different populations become significantly different to each other – this can result in evolutionary selection of individuals with features that help them adapt better to changes in their environment.
- This can result in speciation. **[3]**

13. **Strengths:**

- Conserves whole ecosystems.
- Allows research and education.
- Preserves many habitats and species.
- Prevents hunting and other disturbance from humans.
- Protects many species that are yet to be discovered.

Weaknesses:

- Requires sufficient funding and protection to ensure the area is not disturbed.
- Difficult to establish because of political interference and economic interests.
- Protected area can become an island and lose biodiversity due to edge effects.
- It may be difficult to control outside forces. **[6]**

Conclusion:

Through habitat conservation, the entire ecosystem is supported including individual species within that system. Though its success also depends on making sure sufficient funding exists to support the entire habitat.

[Or any similar type of conclusion to get the final mark.]

Paper 2

Section A

1. (a) organic matter, organisms/nutrients/minerals/air/water **[1]**

(b) • Socio-economic factors – supply and demand, affordability

- Cultural and religious factors – Hindu – no beef, Islam – no pork, Buddhism – vegetarian etc.
- Ecological factors
- Economic factors – money available to spend on food
- Political factors – subsidies and tariffs
- Climate – rainfall, irrigation **[2]**

(c) • Production of corn requires **less** energy/1 kilowatt-hours per pound.

- Production of beef requires **more** energy/32 kilowatt-hours per pound.
- Corn to humans is trophic level 1 to trophic level 2 but beef is trophic level 1 (feed) to cows tropic level 2 to humans trophic level 3 – the longer the food chain the less efficient it is at converting transferring energy.

[You can have any answer here that outlines the 2nd law of thermodynamics] **[2]**

(d)

MEDCs	LEDCs
Mostly large scale commercial farming or for agribusiness	Mostly subsistence farming
Mechanisation/use of fossil fuel	Labour intensive/reliance on manual labour mainly from family

[2]

2. (a) Abiotic **[1]**

(b) • Measure the mass of a sample of soil from a site.

- Dry the sample out until the mass is constant.
- The difference shows the mass of the water evaporated from the sample.
- Repeat for the other two sites. **[2]**

(c) Plants → Insect → Bird → Bird of prey **[1]**

(d) predator-prey / carnivory / predation **[1]**

(e) • The predator benefits by gaining food from prey.

- Its population is stabilized by amount of prey available.
- The prey benefits by predators stabilizing its population.

- Weaker or diseased individuals are more likely to become prey.

- Help to maintaining the gene pool in prey pollution. **[2]**

(f) Birds of prey feed on the coyote which is at trophic level 3 and on the small mammals which are at trophic level 2, so the bird of prey occupies both trophic level 4 for the food chain the coyote is in and tropic level 3 for the food chain the small mammals are in. **[2]**

3. (a) Species diversity – Variety of species per unit area / the number of species and their relative abundance. **[1]**

(b) • H – Habitat destruction and fragmentation – exposes the species to predation and hunting.

- I – Introduced species – brings about competition for food and may outcompete the native species.

- P – Pollution – leads to poisoning of the species resulting in their death.

- P – Practices of agriculture – use of pesticides and fertilizers that poison species.

- O – Overhunting – for sport or trade, reduces the number of organisms. **[2]**

(c) • Flagship species – A species selected due to its appeal to attract public attention/conservation efforts, leading to protection of other species in an area *[Or words to that effect.]*

- Keystone species – A species that is important in the continuation of the functioning of an ecosystem and without which an ecosystem would collapse. **[2]**

(d) 1. Area – Large size is preferable to several small areas.

2. Edge effects – May attract other/exotic species leading to competition and reduction in diversity.

3. Shape – Circle is best because it reduces edge effects.

4. Corridors – Allow gene flow and seasonal movements, reduce collisions between cars and animals.

5. Buffer zones – Should minimize disturbances from outside to the protected area. **[4]**

Paper 2

Section B

4. (a) GHGs such as carbon dioxide, water vapour and methane occur naturally in the atmosphere; they **allow short wavelengths** of the sun's radiation such as ultraviolet (UV) radiation to pass through but **trap longer wavelengths** such as infrared (IR); the trapped radiation creates a thermal blanket that warms the earth maintaining an average temperature that can support life. **[4]**

(b) International organizations are UNEP – Vienna Convention for the Protection of the Ozone Layer and Montreal Protocol on substances that deplete the ozone layer. Both have been successful in getting governments' cooperation in reducing ODSs (Ozone Depleting Substances).

Successes:

- By end of 2002, industrialized countries had reduced ODSs by more than 99%.

- By end of 2002, developing countries had reduced ODSs consumption by more than 50%.

- By 2000, ODSs had been completely phased out in Europe.

Failures:

- Illegal trade in CFCs (chloroflorocarbons) flourished.

- CFCs are persistent, and their effects are felt for a long time. The organizations did not address this.

- HCFCs that replaced CFCs are also ODSs but with lesser impact.

Evaluation: While the world came together to address the problems and there has been a lot of success, especially in industrialized countries, problems still persist, especially in developing countries that are now rapidly industrializing. **[7]**

(c) **Positive feedback**

- Increase in mean global temperatures results from human activities that increase levels of greenhouse gases, the anthropogenic greenhouse effect that leads to climate changes including global warming.

- Increase in mean global temperature through increases in the anthropogenic greenhouse effect leads to less ice in the Arctic. This lowers the lower albedo effect/amount of reflection and leads to increased absorption at the Earth's surface in the Arctic. This leads to increased temperatures in the Arctic plus the increased temperatures from anthropogenic changes to the greenhouse effect accelerate the loss of ice in the Arctic and onwards.

- Increased temperatures lead to melting of permafrost which leads to increased rotting of vegetation under permafrost and increased release of methane. This increases the anthropogenic greenhouse effect which lead to increase in temperature. Increases in temperature lead to loss of permafrost.

Negative feedbacks

- However, reducing human impact on the anthropogenic greenhouse effect could slow, halt or even reverse the loss of ice at the Arctic.

- Removal of carbon dioxide from the atmosphere through increased carbon capture – either in plants or through technology – lowers the anthropogenic greenhouse effect, which would reduce the infra-red radiation reflected back from the atmosphere, which will reduce the global temperature, which will reduce the amount of ice that melts, which will reduce the absorption of heat at the Arctic, which will increase the albedo effect, which will lower the temperature at the Arctic, etc.

- The feedback mechanisms involve long time lags and therefore the models may not be very accurate in their predictions. So, other negative feedbacks that may reduce the melting of ice possibly include: increase in mean global temperature, increased evaporation, higher levels of precipitation, increased snowfall on polar ice caps – all of which result in reduced mean global temperature.

- Higher temperature, more evaporation, more cloud cover, clouds block light from reaching the Earth's surface by reflecting incoming solar radiation. This lowers global temperatures. Less ice melts. **[9]**

5. (a) Simple well annotated diagram or models will really help you to describe your answers for this question.

Succession: Changes to an ecosystem over time.

- The exposure of new land following Glacial retreat in Alaska means that bare rock is left after the retreat of the glacier.

- In time, mosses and lichens start to colonise the rock. As the soils develop, grasses and small herbaceous plants start to grow. These are the pioneer stages characterised by species that are good at benefitting from new opportunities.

- Deeper soils develop and small shrubs colonise these better soils. This is the intermediate stage characterised by longer lived species.

- Eventually trees establish leading to the development of a climax community on mature soils. The climax community is characterised by long lived species and a complex ecosystem.

Zonation: Change in an ecosystem over space. For example, a change in pattern of vegetation up a mountain because of change in an environmental factor such as precipitation.

- Mediterranean vegetation in the Pyrenees follows a zonation with precipitation that is linked to altitude.

- At sea level, a dry grassland and mixed scrub vegetation develops in the dry summer conditions.

- As you increase in altitude, this changes to a scrub woodland as precipitation increases. This changes to closed forest as the amount of precipitation increases with altitude.

- Eventually, at higher altitudes, the woodlands change to alpine grasslands as snow fall and cold temperatures in winter limit the ability of extensive forests to form. **[4]**

(b) • Earth's surface is divided into plates that have moved throughout geological time.

- Plate movements led to creation of both land bridges and physical barriers.

- Physical barriers have led to isolation of gene pools over time. Gene pools change through evolution and can eventually lead to speciation. Speciation leads to increased species diversity.

- Plate movements led to change in environmental conditions. This in turn creates evolutionary pressures that can result natural selection.

- Through chance changes in the gene pool and sexual reproduction, genes that are advantageous to a particular situation can become the dominant genes in the population.

- Natural selection leads to those being best fit to the environment being more successful in passing on their genes. This leads to evolution in the gene pool which can lead to speciation. **[7]**

(c) **Named protected area: Great Barrier Reef**

Things to consider in your answer:

- This is the largest area made by living organisms.

- It has economic values – tourists and the fishing industry.

- It also has cultural and spirituality values.

- There are human activities that threaten the area.

- Degradation of the coral reef due to economic and socio-political pressure.

- Tourism – divers' fins and anchors easily damage the fragile coral.

- Collecting/hunting – damage to coral as tourists break it for souvenirs.

- Harvesting – overfishing which can disrupt the food chain (for example, the unintentional capture of other species while trawling for prawns on the seafloor).

- Anchors used during fishing can also accidentally damage the coral.

- Pollution – run off from agricultural areas into the sea resulting in eutrophication.

- Sewage from coastal settlements further adds to eutrophication.

- Sedimentation (mud pollution) due to deforestation of mangroves.

- Mud pollution makes water cloudy and reduces productivity of the coral reef.

- Pesticides on agricultural land increase pollution further.

Example answer:

Approaches to protecting the Great Barrier Reef include the establishment of marine reserves, with areas that include no-take (no fishing) areas, no-entry areas to boats and encouragement of ecotourism as an alternative to fishing.

Marine reserves have been set up to protect areas of the GBR Such as the GBR Marine Park.

This includes no-take areas where fishing is banned. These allow populations of fish to thrive and reduce the number of species that actively destroy the coral such as the crown-of-thorns starfish. This helps overall ecosystem health and increases biodiversity compared to areas that are still fished. However,

even in no-take areas, poaching can seriously affect the fish population numbers.

Zones with areas that prohibit the entry of boats also help protects species such as the dugong. And contribute to overall biodiversity improvements.

Establishing marine reserves initially affects commercial fishing nearby. However, over time commercial fishing benefits because of protected areas where younger fish could mature and also because of increased economic activity through eco-tourism which has reduced the dependence of local economies on fishing.

However, compared to the money generated through eco-tourism, the amount of money spent on protecting the GBR is small. Also, continued population growth and economic develop along the coast still threaten the GBR.

The biggest threat to the GBR though comes from climate change. Without major success in limiting climate change, the current protection can only slow the loss of biodiversity. So, while there have been successes in protecting the biodiversity of the GBR from human activity, that success is limited and could be threatened in the future. **[9]**

6. (a) (i) • Evaporation from oceans and lakes

- Condensation in the atmosphere

- Precipitation to from clouds

- Freezing **[2]**

(ii) • Ground water

- Lakes

- Oceans

- Glaciers and icesheets

- Atmosphere **[2]**

(b) Humans are supplied with fresh water from the hydrological cycle. This cycle is impacted by human activities, meaning water from underground aquifers is being polluted and withdrawn at a greater rate than it can be replenished. These activities include:

Pollution – meaning water from underground aquifers, rivers lakes and oceans is being polluted.

Deforestation – Fewer trees means less rainfall is trapped, meaning a decrease in interception and infiltration. This also increases surface run off, meaning increased soil erosion as there are fewer roots to bind soil particles. Deforestation also reduces condensation and precipitation, affecting evapotranspiration.

Construction/urbanization – This decreases interception, evapotranspiration and infiltration.

Agriculture – Tilling changes the layout and characteristics of the land, affecting surface run off and infiltration. This changes the movement of surface water to water bodies and reduces the replenishing of ground water.

Irrigation – This can lead to increase in weather such as tornadoes and hail storms, due to salinization.

*[To mark yourself for this question, give yourself a **maximum** of 4 marks for identifying the above impacts (1 mark for each)*

You can then give yourself 4 marks for constructing a well organized system diagram with inputs, processes and outputs. (The example diagram has more inputs than required to gain full marks.)]

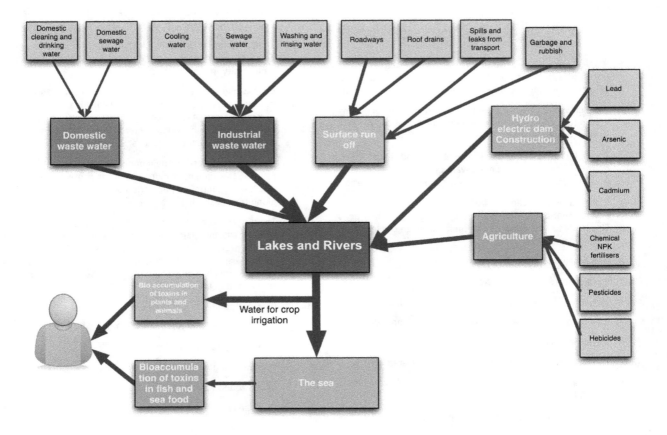

(Source: N Gardner Four Corners Education www.fourcornerseducation.net) **[7]**

(c) *[You need to include ESS terminology in your answer as well as ESS concepts such as climate change, distribution of water supply, population growth, pollution.*

You also need to discuss a range of EVS and link these to specific strategies across the globe.

Below are examples for technocentric and ecocentric EVS. You will need to consider anthropocentric as well.]

Technocentric strategies

Desalination: Allows areas near the sea to have access to fresh water through the removal of salt. However, it often uses fossil fuels to power the process and is expensive. It can also increase the salinity of surrounding marine areas and damage the ecosystem.

Treatment of wastewater: Reuses water, that has already been used, common in developed countries such as the UK. Can be expensive if having to treat polluted waters.

Dams and reservoirs: As well as providing water, can also control floods and generate electricity. Take a large area of land and destroy ecosystems. People are often moved from the land to make way for dams. Can cause additional environmental damage during construction.

[Or similar examples]

Ecocentric strategies

Water conservation: Reduce the use of water already being consumed. Can be difficult to get people to change their habits.

Reuse grey water: Use water from roofs and surfaces for things that do not need fresh water such as toilet flushing. Can be expensive to set up.

Lower population growth in areas where water is scarce, will limit economic development for those areas.

[Or similar examples]

Conclusion

[Write a sound conclusion that weighs the evidence.] For example, `I think that access to water can only be achieved by using a combination of both ecoentric and technocentric strategies because...' **[Max 9]**

SET D

Paper 1

1. Forest

Tropical Rainforest

Equatorial Rainforest **[1]**

2. (a) Any **two** reasons from the list below:

- 2015 has a declining birth (fertility) rate due to increase in female empowerment or equality or education or later marriages whereas 1990 has a high birth (fertility) rate due to less access to education for women.

- 2015 has a declining birth (fertility) rate due to access to contraception or family planning whereas 1990 has higher birth (fertility) rates due to lack of access to contraception or family planning.

- 2015 has lower infant (child) mortality (lower death) rates or higher life expectancy due to improvements or access to medicine or public health whereas 1990 has high death rates (low life expectancy) due to limited access to healthcare.

- 2015 has increasing urbanization or migration from rural areas to cities which results in a lower birth rate whereas 1990 has a high birth rate due to large rural population where children are required to work in agriculture or support parents or religious (cultural) reasons.

- 2015 has introduced population policies or publicity campaigns to reduce fertility rates.

[You must have a reason in your answers to get the marks] **[2]**

(b) Stage 4 **[1]**

3. (a) Any two advantages from:

- Hydroelectric power does not emit smoke/dust/fumes/gases into the atmosphere like fossil fuel powered generators, so reduces pollution.

- Hydroelectric power systems are cheap to maintain once they are set up and established.

- Water is renewable, so it will not run out.

- The dams set up can be used for other purposes like recreation or as a food source. **[2]**

(b) *['Suggest' questions allow you to think of possibilities that may not necessarily be in the Resource Booklet. For example, any two from:]*

- Rivers are found in the south of the country while most people are in the north, so the cost of getting the electricity to people may be more expensive than other sources.
- Fossil fuels are cheaper to exploit, and Venezuela has an abundant supply.
- Technologies for harnessing renewable energy sources are not available on a large scale.
- Inertia within different cultures to changing sources – for example, car culture.
- Traditions in many countries favour non-renewable resources. Locations for renewable energy sources may be limited by the politics of a country. **[2]**

(c) Any one from:

- Flood control, by being able to store flood water behind the dam.
- Fishing may increase with the lake created behind the dam, which may increase employment in the area.
- Recreational activities might develop on the lake.
- The dam provides water irrigation of agriculture in the surrounding area. **[1]**

(d) Any one from:

- Over extraction of water for irrigation.
- Drought over several years. **[1]**

4. (a)
- Most of the rivers are in the south but most people are in the north.
- The supply is contaminated by industrial and agricultural waste
- Changes to rainfall patterns because of El Niño and La Niña that lead to drought.
- Mismanagement of water supply. **[3]**

(b) Eutrophication can occur when rivers, lakes and wetlands have inputs of nutrients such as nitrates and phosphates from agriculture. This increases available nutrients to plants and phytoplankton and results in excess growth, which reduces the amount of oxygen in the water available for other organisms. **[3]**

(c)
- Increased polluted/contaminated water.
- More puddles of stagnant water provide a more breeding ground for mosquitoes.
- Mosquitoes feed on human blood.
- Mosquitoes are vectors for Zika virus. **[2]**

(d) Parasitism is a symbiotic relationship in which one species is benefited and the other species is adversely affected. **[1]**

5. (a) Any two from:

- High rainfall throughout the year
- High level of insolation (sunlight)
- Warm temperatures
- Rainforests contain many niches **[2]**

(b) 3% + 5% + 55% = 63% **[1]**

(c) Product, any one from:

- Fresh water
- Food
- Timber

Service, any one from:

- Regulate biodiversity
- Cycle nutrients

- Provide quality air and climate
- Cultural value **[2]**

6. Any two from:

- Small population
- Reduction in population size
- Specialization (specific diet)
- Poor quality of habitat **[2]**

7.
- Aesthetic: species and habitats are pleasant/beautiful to look at – this may provide inspiration.
- Ecological: endemic species may need specific habitats.
- High biodiversity indicates stability hence species are more likely to survive into the future.
- Healthy ecosystems are more likely to provide other services, such as flood prevention and pollination.
- Disappearance of species can affect the rest of the food chain and the ecosystem. **[4]**

8.
- Venezuela appeared to improve in maternal mortality rates but worsened again after 2005.
- Venezuela has not been successful compared to most other countries, though in Saint Lucia and Jamaica, maternal mortality rates have remained high as well.
- Malaria cases are increasing. Although Venezuela's economy is large and among the best in Latin America, this may have increased malaria through increased mining activity.
- This suggests the country does not appear to prioritize health provision or has reduced its health provision. So, while there has been some success in improving maternal health, these were not sustained and there has been a worsening state in reversing or stopping malaria. **[5]**

Paper 2

Section A

1. (a) (i) Sandy silt loam **[1]**

(ii) Any two from: mineral content; nutrient content; ability to hold water/water holding capacity; drainage; potential to hold organic matter; air spaces. **[2]**

(b) (i) Sand **[1]**

(ii) Any **two** from: Decomposition; Weathering; Nutrient cycling. **[2]**

(c) Roots of trees provide soil with an anchor as well as shelter from wind and rain. When trees are removed this protection is removed and the soil becomes vulnerable to being washed or blown away. **[2]**

2. (a) (i)
- Global warming, which affects climate/weather patterns
- Retreat of polar ice caps and glaciers
- Increase in sea level causing coastal flooding
- Increased flooding **[2]**

(ii) 1.3 – 0.8 = 0.5,

0.5/0.8 x 100 = 62.5%

[You must have the % symbol to get the mark.] **[1]**

(b) Any two from:

- Rocks containing carbon/coal
- Landfill sites
- Manure/sewage
- Extraction/processing/transportation of fossil fuels
- Livestock
- Wetlands/swamps/rice paddies/stagnant water bodies. **[2]**

(ii) Water vapour; ozone **[2]**

(c) Any two from:

- Carbon trading – government-issued permits allowing the emission of carbon dioxide.

- Carbon taxes to encourage producers to reduce emissions; Using biomass as source of fuel.

- Change to renewable energy sources away from fossil fuels.

- Greater use of public transport rather than individual transport.

- Using biomass as source of fuel. **[2]**

3. (a) (i) Secondary succession **[1]**

(ii) Climax community **[1]**

(b) • Long life so are slower growing and late to maturity, so take advantage of the more stable conditions within a later stage of success.

- Fewer, but larger, offspring with a high input of parental care and protection for individual offspring – requires conditions that are more stable.

- Niche specialist so in higher diversity later stages of succession have more niche opportunity.

- Often at higher tropic levels, so require more tropic levels below especially if predators.

[To gain full marks you need two different reasons and the reasons outlined.]

(c)

C (Intermediate)	D (Climax)
GPP lower due to fewer smaller plant species	GPP higher due to more larger plant species
NPP higher as more GPP is converted to new growth	NPP lower as a lot of GPP is used in maintaining the organisms through respiration
Biomass lower because of smaller plant forms	Biomass higher because of larger plant forms with a lot of stored biomass

[3]

Paper 2

Section B

4. (a) (i) Habitat conservation: Practices that aim to conserve, protect and restore habitats. Successful habitat protect has occurred in the Ujung Kulon National Park in Indonesia, where, by protecting the habitat, the last remaining population of Javan Rhinoceros has grown.

Species conservation: Conservation of species in their natural habitat (in-situ conservation) or outside their natural habitat (ex-situ conservation). Such as ex-situ gorilla breeding programs at Jersey Zoo on the Channel Islands. **[4]**

(b) BMWP sampling – works by examining the composition of a collected sample of macroinvertebrates. The greater the number of species that require well oxygenated, unpolluted water, the better the water quality.

Kick net – is used to capture small aquatic macroinvertebrates. It is placed on the river bed and the area in front of the river bed is disturbed (kicked) for 10 seconds so that macroinvertebrates in the sand and gravels get caught in the net.

Any macroinvertebrates caught in the net are identified to the family level, this can be done with the live macroinvertebrates that can be returned to the ecosystem after sampling. The macroinvertebrates found are recorded and scored against the Biological Monitoring Working Party scores for each family. The higher the score the less resistant to low oxygen levels and pollution a family is.

The total score is summed up and the higher the BMWP score the better the water quality. This can be used to compare different aquatic systems or to compare different parts of the same aquatic system such as a river to look for evidence of pollution events that may not be picked up by chemical analysis where the pollutant may have already left the system. **[7]**

(c) Between 1980 and 2000, the increase in human population increased the need for agriculture, energy and infrastructure development. Tropical rainforests are home to a large number of species and deforestation leads to the loss of these species (plants, animals).

Growth of consumerism created a need for products made from wood, often taken from tropical forests.

Since 2000, an increasing awareness in the value of tropical forests as life-support systems that provide oxygen and mediate climates, has led to an increase in protection for areas of the forests. Increased international agreements about reducing deforestation have been implemented and NGO have been successful in changing the way people think about products from tropical forests.

International conventions such as Kyoto Protocol, Paris Agreement, UN Convention on Climate Change, carbon trading, REDD (Reduced Emissions from Deforestation & Degradation) have helped focus on the role of retaining forest cover. To reduce impacts of water systems, climate, flood defense, etc.

In the MEDCs, a growing demand for meat with smaller carbon footprints as reduced imports of meat from areas where tropical forests have been cleared for beef farming. Similarly, with Palm oil.

Changes away from reliance on wood as fuel in LEDCs, to both non-renewable fossil fuel and renewable fuels, has reduced the need for charcoal, etc. **[9]**

5. (a) Primary pollutants are pollutants emitted directly from a source that have an effect in the form they are in.

Secondary pollutants are not directly emitted as such, but form when other pollutants (primary pollutants) react or when a pollutant changes over time.

Examples of primary pollutants:

- Sulfur dioxide

- Nitrogen oxide

- Nitrogen dioxide

- Carbon monoxide

- Volatile organic compounds

- Particulate matter

- Mercury

Examples of secondary pollutants:

- Ozone

- Nitric acid

- Sulfuric acid to acid rain

[You need the distinction between each and an example of each.]
[4]

(b) Burning fossil fuels releases nitrogen monoxide and hydrocarbons; nitrogen monoxide (NO) reacts with oxygen to form nitrogen dioxide (NO_2). NO_2 absorbs sunlight and breaks down to form oxygen atoms. These combine with oxygen in the air to form ozone, a secondary pollutant.

Photochemical reactions between nitrogen oxides and volatile organic compounds (VOCs) from fuel combustion take several hours to produce ozone. Ozone concentration is greatest in the early afternoon when there is more sunlight and temperatures are warm.

Ozone reacts with the primary pollutants in the presence of sunlight to form photochemical smog. **[7]**

(c) Alter the human activities that produce acid rain:

- Use alternatives to fossil fuels.

- Use renewable energy sources for electricity.

- Reduce overall demand for electricity by running education campaigns for the population, e.g. turning off lights.

- Use less private transport – walk, cycle, car pool/sharing, take public transport.

- Use low sulfur fuels – remove sulfur before burning.

- Economy is reliant on fossil fuels – in many MEDC, cars remain the primary form of transport.

- Demand for power is increasing as countries industrialize.

- Difficult to get people to change what they have been doing and often requires legislation or promotion to make the change occur. Often requires infrastructure changes to make it possible for a lot of people to make the change.

- As the cost of renewable energy decreases, more people are persuaded of its value.

Regulate and reduce the pollutants that cause acid rain:

- Install cleaning technologies at the point of emission, e.g. scrubbing chimneys to remove sulfur dioxide.

- Install catalytic converters to convert nitrogen oxides to nitrogen gas.

- Expensive: cost is passed on to consumers.

- Catalysers are expensive to buy but cost-effective if maintained well.

- If costs rise beyond economic viability the possibility that the pollution causing industries are exported from MEDCs to LEDCs where regulations are less tight.

- To be effective globally, requires international agreement that may take a long time to develop and may not be as complete as needed (COP26).

[Your answer requires an evaluation of the strengths and weaknesses of each, not just a description.] **[9]**

6. (a) Market value of natural capital changes over time and varies across regions.

Status of a resource is influenced by cultural, economic and technological factors.

For example: uranium, before the development of nuclear power uranium was not considered as natural capital even though it was in the Earth's mineral deposits; nuclear power development changed its status to an expensive resource. **[4]**

(b) Renewable natural capital can be generated/replaced at the same rate at which it is being used; if used beyond its natural income, this use becomes unsustainable.

For example, groundwater gets contaminated by agricultural products such as fertilizers, pesticides, herbicides; contamination by run-off from storage tanks, septic tanks and infiltration from landfills.

Over-extraction for domestic, agricultural and industrial use, lowering the water tables; allowing salt water along coasts to contaminate aquifers.

Reduced water supply available for agriculture results in reduced yields and increased cost of water for industry and agriculture; This affects the economy negatively. **[7]**

(c) Carrying capacity is the maximum number of species/load that a given area can support sustainably.

It is difficult to determine carrying capacity for humans because:

- Humans use a wide range of resources.

- When one resource is in short supply, humans can substitute it for another. So, humans can use wood, metal, bricks, concrete to provide shelter.

- Lifestyles affect which resources are required. Lifestyles differ from time to time and between populations so the needs of one population may not be the same as the needs of another population.

- Development in technology changes the resources required and available for human use so humans can recycle or import resources. Importing resources means that humans can go beyond the boundaries of their local resources. This increases the carrying capacity.

- Importing resources increases the carrying capacity of the local population but not the global carrying capacity.

Problems of carrying capacity and ecological footprint:

Because of this it is difficult to estimate the carrying capacity of the planet for humans as it needs to consider a complex set of measures that include ecological footprint. The effect of humans on the environment not only in terms of what we need but also what we degrade. If we degrade the planet's natural resources faster than they can recover that will in itself limit the carrying capacity.

Measuring the carrying capacity now and projecting future models from it may not match the realities of the future. New technology may be developed that increase the CC and reduce our ecological footprint, but also, we may degrade planet and the CC faster than we model if we reach an unforeseen tipping point.

Ecological footprint is the inverse of carrying capacity as it measures how much is used as well as how much waste is produced, but it may be difficult to estimate CC from ecological footprint, even though both depend on how fast resources are used. However, ecological footprint also includes a measure of waste produced so it may decrease carrying capacity as more resources are used and waste adds to pollution.

[You need to include the concept of and problems of ecological footprint within your answer to develop a balanced discussion.] **[9]**

CPSIA information can be obtained
at www.ICGtesting.com
Printed in the USA
LVHW021128020723
751358LV00007B/612

THIS BOOK BELONGS TO:

CONTACT INFORMATION	
NAME:	
ADDRESS:	
PHONE:	

START / END DATES

_____ / _____ / _____ TO _____ / _____ / _____

DEDICATION

This book is dedicated to all the amazing goat owners who love raising and taking care of GOATS!

You are my inspiration in producing books and I'm excited to help in the planning of GOAT CARE into your day around the world!

How to Use this Goat Record Keeping Notebook:

The purpose of this Goat Raising Log Book is for anyone is to keep all various goat breeding activities and information organized in one easy to find spot.

Here are some simple guidelines to follow so you can make the most of using this book:

1. The first "Goat Information" section is for you to write out your goat's name, tattoo, breed, physical characteristics and a pedigree chart so you can track your goat raising adventures.

2. Most ideas are inspired by something we have seen. Use the "Medical Information" section to write down the date, nature of any illness, parasite control, testing record and a vaccination record so you can go back there to be reminded later.

3. The "Doe's Kidding Record" section is for you to write out THAT Doe's name, date, breed, kidding date and any important information for each Doe.

4. Some ideas require listing them out, the "Buck;s Record Of Progeny" section is great for using this to record the year, bred to, kids and much more for you to use to keep track and refer back to this list later.

5. The "Goat Record" section is so you can list out the Goat's name, breed, Identification, date of birth, weaning date, and month to month weight tracker... and be inspired to add to your goat record.

6. And finally pages with a "Milk Production" section for you to make entries about month to month milk produced, yearly production and the value per lbs... and much much more.

Have fun!

GOAT INFORMATION

PHOTO

NAME		☐ BUCK	☐ DOE
BREED		BIRTH DATE:	
DATE ACQUIRED:	HOW ACQUIRED: ☐ BORN ON FARM ☐ PURCHASED ☐ LEASED		
COLORS / IDENTIFYING MARKS:			
PURPOSE: ☐ MILK ☐ MEAT ☐ PET ☐ OTHER			

PEDIGREE CHART

- SIRE
  - GRAND SIRE
    - GREAT GRAND SIRE
    - GREAT GRAND DAM
  - GRAND DAM
    - GREAT GRAND SIRE
    - GREAT GRAND DAM
- DAM
  - GRAND SIRE
    - GREAT GRAND SIRE
    - GREAT GRAND DAM
  - GRAND DAM
    - GREAT GRAND SIRE
    - GREAT GRAND DAM

MEDICAL INFORMATION

INJURY OR ILLNESS

DATE	DESCRIPTION OR NATURE OF ILLNESS	TREATMENT

PARASITE CONTROL

DATE	METHOD OR DEWORMER	DATE	METHOD OR DEWORMER

TESTING RECORD

DATE	TEST PERFORMED (CAE, CL, TB...)	RESULT	DATE	TEST PERFORMED (CAE, CL, TB...)	RESULT

INJURY OR ILLNESS

DATE	TARGET DISEASE	DRUG OR SUPPLEMENT USED	DOSAGE	RESULTS

DOE'S KIDDING RECORD

DOE'S NAME:	

DATE BREED	KIDDING DATE	# OF KIDS	SEX D/B	NAME OF KID	SIRE OF KID	WEIGHT	TATTOO

BUCK'S RECORD OF PROGENY

DOE'S NAME:			

YEAR	BRED TO	KIDS	DOE/BUCK

GOAT RECORD

GOAT'S NAME:		IDENTIFICATION:	
BREED:	DATE OF BIRTH:	DATE OF WEANED:	

WEIGHT (POUNDS)

BIRTH	JAN	FEB	MAR	APR	MAY	JUN	JUL	AUG	SEPT	OCT	NOV	DEC	FINAL

FEED RECORD

	JAN	FEB	MAR	APR	MAY	JUN	JUL	AUG	SEPT	OCT	NOV	DEC	TOTAL
GRAIN													
GRAIN													
PASTURE													

MILK PRODUCTION

GOAT'S NAME:		IDENTIFICATION:	
BREED:	DATE OF BIRTH:	KIDDING DATE:	

JANUARY		AVERAGE LBS / DAY X 31 DAYS =		LBS
FEBRUARY		AVERAGE LBS / DAY X 31 DAYS =		LBS
MARCH		AVERAGE LBS / DAY X 31 DAYS =		LBS
APRIL		AVERAGE LBS / DAY X 31 DAYS =		LBS
MAY		AVERAGE LBS / DAY X 31 DAYS =		LBS
JUNE		AVERAGE LBS / DAY X 31 DAYS =		LBS
JULY		AVERAGE LBS / DAY X 31 DAYS =		LBS
AUGUST		AVERAGE LBS / DAY X 31 DAYS =		LBS
SEPTEMBER		AVERAGE LBS / DAY X 31 DAYS =		LBS
OCTOBER		AVERAGE LBS / DAY X 31 DAYS =		LBS
NOVEMBER		AVERAGE LBS / DAY X 31 DAYS =		LBS
DECEMBER		AVERAGE LBS / DAY X 31 DAYS =		LBS
YEARLY TOTAL MILK PRODUCED =				LBS
TOTAL VALUE OF MILK PRODUCED FOR THE YEAR				
	LBS X $		VALUE PER LBS =	

GOAT INFORMATION

PHOTO

NAME	☐ BUCK	☐ DOE
BREED	BIRTH DATE:	
DATE ACQUIRED:	HOW ACQUIRED: ☐ BORN ON FARM ☐ PURCHASED ☐ LEASED	
COLORS / IDENTIFYING MARKS:		
PURPOSE: ☐ MILK ☐ MEAT ☐ PET ☐ OTHER		

PEDIGREE CHART

- SIRE
  - GRAND SIRE
    - GREAT GRAND SIRE
    - GREAT GRAND DAM
  - GRAND DAM
    - GREAT GRAND SIRE
    - GREAT GRAND DAM
- DAM
  - GRAND SIRE
    - GREAT GRAND SIRE
    - GREAT GRAND DAM
  - GRAND DAM
    - GREAT GRAND SIRE
    - GREAT GRAND DAM

MEDICAL INFORMATION

INJURY OR ILLNESS

DATE	DESCRIPTION OR NATURE OF ILLNESS	TREATMENT

PARASITE CONTROL

DATE	METHOD OR DEWORMER	DATE	METHOD OR DEWORMER

TESTING RECORD

DATE	TEST PERFORMED (CAE, CL, TB...)	RESULT	DATE	TEST PERFORMED (CAE, CL, TB...)	RESULT

INJURY OR ILLNESS

DATE	TARGET DISEASE	DRUG OR SUPPLEMENT USED	DOSAGE	RESULTS

DOE'S KIDDING RECORD

DOE'S NAME:

DATE BREED	KIDDING DATE	# OF KIDS	SEX D/B	NAME OF KID	SIRE OF KID	WEIGHT	TATTOO

BUCK'S RECORD OF PROGENY

DOE'S NAME:	

YEAR	BRED TO	KIDS	DOE/BUCK

GOAT RECORD

GOAT'S NAME:		IDENTIFICATION:	
BREED:	DATE OF BIRTH:		DATE OF WEANED:

WEIGHT (POUNDS)

BIRTH	JAN	FEB	MAR	APR	MAY	JUN	JUL	AUG	SEPT	OCT	NOV	DEC	FINAL

FEED RECORD

	JAN	FEB	MAR	APR	MAY	JUN	JUL	AUG	SEPT	OCT	NOV	DEC	TOTAL
GRAIN													
GRAIN													
PASTURE													

MILK PRODUCTION

GOAT'S NAME:		IDENTIFICATION:	
BREED:	DATE OF BIRTH:	KIDDING DATE:	

JANUARY		AVERAGE LBS / DAY X 31 DAYS =		LBS
FEBRUARY		AVERAGE LBS / DAY X 31 DAYS =		LBS
MARCH		AVERAGE LBS / DAY X 31 DAYS =		LBS
APRIL		AVERAGE LBS / DAY X 31 DAYS =		LBS
MAY		AVERAGE LBS / DAY X 31 DAYS =		LBS
JUNE		AVERAGE LBS / DAY X 31 DAYS =		LBS
JULY		AVERAGE LBS / DAY X 31 DAYS =		LBS
AUGUST		AVERAGE LBS / DAY X 31 DAYS =		LBS
SEPTEMBER		AVERAGE LBS / DAY X 31 DAYS =		LBS
OCTOBER		AVERAGE LBS / DAY X 31 DAYS =		LBS
NOVEMBER		AVERAGE LBS / DAY X 31 DAYS =		LBS
DECEMBER		AVERAGE LBS / DAY X 31 DAYS =		LBS
YEARLY TOTAL MILK PRODUCED =				LBS

TOTAL VALUE OF MILK PRODUCED FOR THE YEAR

	LBS X $		VALUE PER LBS =	

GOAT INFORMATION

PHOTO

NAME		☐ BUCK	☐ DOE
BREED		BIRTH DATE:	
DATE ACQUIRED:	HOW ACQUIRED: ☐ BORN ON FARM ☐ PURCHASED ☐ LEASED		
COLORS / IDENTIFYING MARKS:			
PURPOSE: ☐ MILK ☐ MEAT ☐ PET ☐ OTHER			

PEDIGREE CHART

SIRE

GRAND SIRE

GREAT GRAND SIRE

GREAT GRAND DAM

GRAND DAM

GREAT GRAND SIRE

GREAT GRAND DAM

DAM

GRAND SIRE

GREAT GRAND SIRE

GREAT GRAND DAM

GRAND DAM

GREAT GRAND SIRE

GREAT GRAND DAM

MEDICAL INFORMATION

INJURY OR ILLNESS

DATE	DESCRIPTION OR NATURE OF ILLNESS	TREATMENT

PARASITE CONTROL

DATE	METHOD OR DEWORMER	DATE	METHOD OR DEWORMER

TESTING RECORD

DATE	TEST PERFORMED (CAE, CL, TB...)	RESULT	DATE	TEST PERFORMED (CAE, CL, TB...)	RESULT

INJURY OR ILLNESS

DATE	TARGET DISEASE	DRUG OR SUPPLEMENT USED	DOSAGE	RESULTS

DOE'S KIDDING RECORD

DOE'S NAME:

DATE BREED	KIDDING DATE	# OF KIDS	SEX D/B	NAME OF KID	SIRE OF KID	WEIGHT	TATTOO

BUCK'S RECORD OF PROGENY

DOE'S NAME:	

YEAR	BRED TO	KIDS	DOE/BUCK

GOAT RECORD

GOAT'S NAME:		IDENTIFICATION:
BREED:	DATE OF BIRTH:	DATE OF WEANED:

WEIGHT (POUNDS)

BIRTH	JAN	FEB	MAR	APR	MAY	JUN	JUL	AUG	SEPT	OCT	NOV	DEC	FINAL

FEED RECORD

	JAN	FEB	MAR	APR	MAY	JUN	JUL	AUG	SEPT	OCT	NOV	DEC	TOTAL
GRAIN													
GRAIN													
PASTURE													

MILK PRODUCTION

GOAT'S NAME:		IDENTIFICATION:	
BREED:	DATE OF BIRTH:	KIDDING DATE:	

JANUARY		AVERAGE LBS / DAY X 31 DAYS =		LBS
FEBRUARY		AVERAGE LBS / DAY X 31 DAYS =		LBS
MARCH		AVERAGE LBS / DAY X 31 DAYS =		LBS
APRIL		AVERAGE LBS / DAY X 31 DAYS =		LBS
MAY		AVERAGE LBS / DAY X 31 DAYS =		LBS
JUNE		AVERAGE LBS / DAY X 31 DAYS =		LBS
JULY		AVERAGE LBS / DAY X 31 DAYS =		LBS
AUGUST		AVERAGE LBS / DAY X 31 DAYS =		LBS
SEPTEMBER		AVERAGE LBS / DAY X 31 DAYS =		LBS
OCTOBER		AVERAGE LBS / DAY X 31 DAYS =		LBS
NOVEMBER		AVERAGE LBS / DAY X 31 DAYS =		LBS
DECEMBER		AVERAGE LBS / DAY X 31 DAYS =		LBS
YEARLY TOTAL MILK PRODUCED =				LBS

TOTAL VALUE OF MILK PRODUCED FOR THE YEAR

	LBS X $		VALUE PER LBS =	

GOAT INFORMATION

PHOTO

NAME		☐ BUCK	☐ DOE
BREED		BIRTH DATE:	
DATE ACQUIRED:	HOW ACQUIRED: ☐ BORN ON FARM ☐ PURCHASED ☐ LEASED		
COLORS / IDENTIFYING MARKS:			
PURPOSE: ☐ MILK ☐ MEAT ☐ PET ☐ OTHER			

PEDIGREE CHART

SIRE

GRAND SIRE

GREAT GRAND SIRE

GREAT GRAND DAM

GRAND DAM

GREAT GRAND SIRE

GREAT GRAND DAM

DAM

GRAND SIRE

GREAT GRAND SIRE

GREAT GRAND DAM

GRAND DAM

GREAT GRAND SIRE

GREAT GRAND DAM

MEDICAL INFORMATION

INJURY OR ILLNESS

DATE	DESCRIPTION OR NATURE OF ILLNESS	TREATMENT

PARASITE CONTROL

DATE	METHOD OR DEWORMER	DATE	METHOD OR DEWORMER

TESTING RECORD

DATE	TEST PERFORMED (CAE, CL, TB...)	RESULT	DATE	TEST PERFORMED (CAE, CL, TB...)	RESULT

INJURY OR ILLNESS

DATE	TARGET DISEASE	DRUG OR SUPPLEMENT USED	DOSAGE	RESULTS

DOE'S KIDDING RECORD

DOE'S NAME:	

DATE BREED	KIDDING DATE	# OF KIDS	SEX D/B	NAME OF KID	SIRE OF KID	WEIGHT	TATTOO

BUCK'S RECORD OF PROGENY

DOE'S NAME:	

YEAR	BRED TO	KIDS	DOE/BUCK

GOAT RECORD

GOAT'S NAME: | IDENTIFICATION:

BREED: | DATE OF BIRTH: | DATE OF WEANED:

WEIGHT (POUNDS)													
BIRTH	JAN	FEB	MAR	APR	MAY	JUN	JUL	AUG	SEPT	OCT	NOV	DEC	FINAL

FEED RECORD												
	JAN	FEB	MAR	APR	MAY	JUN	JUL	AUG	SEPT	OCT	NOV	DEC	TOTAL
GRAIN													
GRAIN													
PASTURE													

MILK PRODUCTION

GOAT'S NAME:		IDENTIFICATION:		
BREED:		DATE OF BIRTH:		KIDDING DATE:

JANUARY		AVERAGE LBS / DAY X 31 DAYS =		LBS
FEBRUARY		AVERAGE LBS / DAY X 31 DAYS =		LBS
MARCH		AVERAGE LBS / DAY X 31 DAYS =		LBS
APRIL		AVERAGE LBS / DAY X 31 DAYS =		LBS
MAY		AVERAGE LBS / DAY X 31 DAYS =		LBS
JUNE		AVERAGE LBS / DAY X 31 DAYS =		LBS
JULY		AVERAGE LBS / DAY X 31 DAYS =		LBS
AUGUST		AVERAGE LBS / DAY X 31 DAYS =		LBS
SEPTEMBER		AVERAGE LBS / DAY X 31 DAYS =		LBS
OCTOBER		AVERAGE LBS / DAY X 31 DAYS =		LBS
NOVEMBER		AVERAGE LBS / DAY X 31 DAYS =		LBS
DECEMBER		AVERAGE LBS / DAY X 31 DAYS =		LBS
YEARLY TOTAL MILK PRODUCED =				LBS
TOTAL VALUE OF MILK PRODUCED FOR THE YEAR				
	LBS X $		VALUE PER LBS =	

GOAT INFORMATION

PHOTO

NAME	☐ BUCK	☐ DOE
BREED	BIRTH DATE:	

DATE ACQUIRED:	HOW ACQUIRED: ☐ BORN ON FARM ☐ PURCHASED ☐ LEASED

COLORS / IDENTIFYING MARKS:

PURPOSE:	☐ MILK	☐ MEAT	☐ PET	☐ OTHER

PEDIGREE CHART

- SIRE
  - GRAND SIRE
    - GREAT GRAND SIRE
    - GREAT GRAND DAM
  - GRAND DAM
    - GREAT GRAND SIRE
    - GREAT GRAND DAM
- DAM
  - GRAND SIRE
    - GREAT GRAND SIRE
    - GREAT GRAND DAM
  - GRAND DAM
    - GREAT GRAND SIRE
    - GREAT GRAND DAM

MEDICAL INFORMATION

INJURY OR ILLNESS

DATE	DESCRIPTION OR NATURE OF ILLNESS	TREATMENT

PARASITE CONTROL

DATE	METHOD OR DEWORMER	DATE	METHOD OR DEWORMER

TESTING RECORD

DATE	TEST PERFORMED (CAE, CL, TB...)	RESULT	DATE	TEST PERFORMED (CAE, CL, TB...)	RESULT

INJURY OR ILLNESS

DATE	TARGET DISEASE	DRUG OR SUPPLEMENT USED	DOSAGE	RESULTS

DOE'S KIDDING RECORD

DOE'S NAME:

DATE BREED	KIDDING DATE	# OF KIDS	SEX D/B	NAME OF KID	SIRE OF KID	WEIGHT	TATTOO

BUCK'S RECORD OF PROGENY

DOE'S NAME:	

YEAR	BRED TO	KIDS	DOE/BUCK

GOAT RECORD

GOAT'S NAME:		IDENTIFICATION:	
BREED:	DATE OF BIRTH:	DATE OF WEANED:	

WEIGHT (POUNDS)

BIRTH	JAN	FEB	MAR	APR	MAY	JUN	JUL	AUG	SEPT	OCT	NOV	DEC	FINAL

FEED RECORD

	JAN	FEB	MAR	APR	MAY	JUN	JUL	AUG	SEPT	OCT	NOV	DEC	TOTAL
GRAIN													
GRAIN													
PASTURE													

MILK PRODUCTION

GOAT'S NAME:

IDENTIFICATION:

BREED:

DATE OF BIRTH:

KIDDING DATE:

JANUARY		AVERAGE LBS / DAY X 31 DAYS =		LBS
FEBRUARY		AVERAGE LBS / DAY X 31 DAYS =		LBS
MARCH		AVERAGE LBS / DAY X 31 DAYS =		LBS
APRIL		AVERAGE LBS / DAY X 31 DAYS =		LBS
MAY		AVERAGE LBS / DAY X 31 DAYS =		LBS
JUNE		AVERAGE LBS / DAY X 31 DAYS =		LBS
JULY		AVERAGE LBS / DAY X 31 DAYS =		LBS
AUGUST		AVERAGE LBS / DAY X 31 DAYS =		LBS
SEPTEMBER		AVERAGE LBS / DAY X 31 DAYS =		LBS
OCTOBER		AVERAGE LBS / DAY X 31 DAYS =		LBS
NOVEMBER		AVERAGE LBS / DAY X 31 DAYS =		LBS
DECEMBER		AVERAGE LBS / DAY X 31 DAYS =		LBS
YEARLY TOTAL MILK PRODUCED =				LBS

TOTAL VALUE OF MILK PRODUCED FOR THE YEAR

	LBS X $		VALUE PER LBS =	

GOAT INFORMATION

PHOTO

NAME	☐ BUCK	☐ DOE
BREED	BIRTH DATE:	
DATE ACQUIRED:	HOW ACQUIRED: ☐ BORN ON FARM ☐ PURCHASED ☐ LEASED	
COLORS / IDENTIFYING MARKS:		
PURPOSE: ☐ MILK ☐ MEAT ☐ PET ☐ OTHER		

PEDIGREE CHART

SIRE

DAM

GRAND SIRE

GRAND DAM

GRAND SIRE

GRAND DAM

GREAT GRAND SIRE

GREAT GRAND DAM

GREAT GRAND SIRE

GREAT GRAND DAM

GREAT GRAND SIRE

GREAT GRAND DAM

GREAT GRAND SIRE

GREAT GRAND DAM

MEDICAL INFORMATION

INJURY OR ILLNESS

DATE	DESCRIPTION OR NATURE OF ILLNESS	TREATMENT

PARASITE CONTROL

DATE	METHOD OR DEWORMER	DATE	METHOD OR DEWORMER

TESTING RECORD

DATE	TEST PERFORMED (CAE, CL, TB...)	RESULT	DATE	TEST PERFORMED (CAE, CL, TB...)	RESULT

INJURY OR ILLNESS

DATE	TARGET DISEASE	DRUG OR SUPPLEMENT USED	DOSAGE	RESULTS

DOE'S KIDDING RECORD

DOE'S NAME:	

DATE BREED	KIDDING DATE	# OF KIDS	SEX D/B	NAME OF KID	SIRE OF KID	WEIGHT	TATTOO

BUCK'S RECORD OF PROGENY

DOE'S NAME:

YEAR	BRED TO	KIDS	DOE/BUCK

GOAT RECORD

GOAT'S NAME:		IDENTIFICATION:	
BREED:	DATE OF BIRTH:	DATE OF WEANED:	

WEIGHT (POUNDS)

BIRTH	JAN	FEB	MAR	APR	MAY	JUN	JUL	AUG	SEPT	OCT	NOV	DEC	FINAL

FEED RECORD

	JAN	FEB	MAR	APR	MAY	JUN	JUL	AUG	SEPT	OCT	NOV	DEC	TOTAL
GRAIN													
GRAIN													
PASTURE													

MILK PRODUCTION

GOAT'S NAME:

IDENTIFICATION:

BREED:

DATE OF BIRTH:

KIDDING DATE:

JANUARY		AVERAGE LBS / DAY X 31 DAYS =		LBS
FEBRUARY		AVERAGE LBS / DAY X 31 DAYS =		LBS
MARCH		AVERAGE LBS / DAY X 31 DAYS =		LBS
APRIL		AVERAGE LBS / DAY X 31 DAYS =		LBS
MAY		AVERAGE LBS / DAY X 31 DAYS =		LBS
JUNE		AVERAGE LBS / DAY X 31 DAYS =		LBS
JULY		AVERAGE LBS / DAY X 31 DAYS =		LBS
AUGUST		AVERAGE LBS / DAY X 31 DAYS =		LBS
SEPTEMBER		AVERAGE LBS / DAY X 31 DAYS =		LBS
OCTOBER		AVERAGE LBS / DAY X 31 DAYS =		LBS
NOVEMBER		AVERAGE LBS / DAY X 31 DAYS =		LBS
DECEMBER		AVERAGE LBS / DAY X 31 DAYS =		LBS
YEARLY TOTAL MILK PRODUCED =				LBS

TOTAL VALUE OF MILK PRODUCED FOR THE YEAR

	LBS X $		VALUE PER LBS =	

GOAT INFORMATION

PHOTO

NAME		☐ BUCK	☐ DOE
BREED		BIRTH DATE:	
DATE ACQUIRED:	HOW ACQUIRED: ☐ BORN ON FARM ☐ PURCHASED ☐ LEASED		
COLORS / IDENTIFYING MARKS:			
PURPOSE: ☐ MILK ☐ MEAT ☐ PET ☐ OTHER			

PEDIGREE CHART

			GREAT GRAND SIRE
		GRAND SIRE	
			GREAT GRAND DAM
	SIRE		GREAT GRAND SIRE
		GRAND DAM	
			GREAT GRAND DAM
			GREAT GRAND SIRE
		GRAND SIRE	
			GREAT GRAND DAM
	DAM		GREAT GRAND SIRE
		GRAND DAM	
			GREAT GRAND DAM

MEDICAL INFORMATION

INJURY OR ILLNESS

DATE	DESCRIPTION OR NATURE OF ILLNESS	TREATMENT

PARASITE CONTROL

DATE	METHOD OR DEWORMER	DATE	METHOD OR DEWORMER

TESTING RECORD

DATE	TEST PERFORMED (CAE, CL, TB...)	RESULT	DATE	TEST PERFORMED (CAE, CL, TB...)	RESULT

INJURY OR ILLNESS

DATE	TARGET DISEASE	DRUG OR SUPPLEMENT USED	DOSAGE	RESULTS

DOE'S KIDDING RECORD

DOE'S NAME:

DATE BREED	KIDDING DATE	# OF KIDS	SEX D/B	NAME OF KID	SIRE OF KID	WEIGHT	TATTOO

BUCK'S RECORD OF PROGENY

DOE'S NAME:	

YEAR	BRED TO	KIDS	DOE/BUCK

GOAT RECORD

GOAT'S NAME:		IDENTIFICATION:	
BREED:	DATE OF BIRTH:	DATE OF WEANED:	

WEIGHT (POUNDS)

BIRTH	JAN	FEB	MAR	APR	MAY	JUN	JUL	AUG	SEPT	OCT	NOV	DEC	FINAL

FEED RECORD

	JAN	FEB	MAR	APR	MAY	JUN	JUL	AUG	SEPT	OCT	NOV	DEC	TOTAL
GRAIN													
GRAIN													
PASTURE													

MILK PRODUCTION

GOAT'S NAME:		IDENTIFICATION:		
BREED:		DATE OF BIRTH:	KIDDING DATE:	

JANUARY		AVERAGE LBS / DAY X 31 DAYS =		LBS
FEBRUARY		AVERAGE LBS / DAY X 31 DAYS =		LBS
MARCH		AVERAGE LBS / DAY X 31 DAYS =		LBS
APRIL		AVERAGE LBS / DAY X 31 DAYS =		LBS
MAY		AVERAGE LBS / DAY X 31 DAYS =		LBS
JUNE		AVERAGE LBS / DAY X 31 DAYS =		LBS
JULY		AVERAGE LBS / DAY X 31 DAYS =		LBS
AUGUST		AVERAGE LBS / DAY X 31 DAYS =		LBS
SEPTEMBER		AVERAGE LBS / DAY X 31 DAYS =		LBS
OCTOBER		AVERAGE LBS / DAY X 31 DAYS =		LBS
NOVEMBER		AVERAGE LBS / DAY X 31 DAYS =		LBS
DECEMBER		AVERAGE LBS / DAY X 31 DAYS =		LBS
YEARLY TOTAL MILK PRODUCED =				LBS

TOTAL VALUE OF MILK PRODUCED FOR THE YEAR

	LBS X $		VALUE PER LBS =	

GOAT INFORMATION

PHOTO

NAME	☐ BUCK	☐ DOE
BREED	BIRTH DATE:	
DATE ACQUIRED:	HOW ACQUIRED: ☐ BORN ON FARM ☐ PURCHASED ☐ LEASED	
COLORS / IDENTIFYING MARKS:		
PURPOSE: ☐ MILK ☐ MEAT ☐ PET ☐ OTHER		

PEDIGREE CHART

- SIRE
  - GRAND SIRE
    - GREAT GRAND SIRE
    - GREAT GRAND DAM
  - GRAND DAM
    - GREAT GRAND SIRE
    - GREAT GRAND DAM
- DAM
  - GRAND SIRE
    - GREAT GRAND SIRE
    - GREAT GRAND DAM
  - GRAND DAM
    - GREAT GRAND SIRE
    - GREAT GRAND DAM

MEDICAL INFORMATION

INJURY OR ILLNESS

DATE	DESCRIPTION OR NATURE OF ILLNESS	TREATMENT

PARASITE CONTROL

DATE	METHOD OR DEWORMER	DATE	METHOD OR DEWORMER

TESTING RECORD

DATE	TEST PERFORMED (CAE, CL, TB...)	RESULT	DATE	TEST PERFORMED (CAE, CL, TB...)	RESULT

INJURY OR ILLNESS

DATE	TARGET DISEASE	DRUG OR SUPPLEMENT USED	DOSAGE	RESULTS

DOE'S KIDDING RECORD

DOE'S NAME:

DATE BREED	KIDDING DATE	# OF KIDS	SEX D/B	NAME OF KID	SIRE OF KID	WEIGHT	TATTOO

BUCK'S RECORD OF PROGENY

DOE'S NAME:	

YEAR	BRED TO	KIDS	DOE/BUCK

GOAT RECORD

GOAT'S NAME:		IDENTIFICATION:	
BREED:	DATE OF BIRTH:	DATE OF WEANED:	

WEIGHT (POUNDS)

BIRTH	JAN	FEB	MAR	APR	MAY	JUN	JUL	AUG	SEPT	OCT	NOV	DEC	FINAL

FEED RECORD

	JAN	FEB	MAR	APR	MAY	JUN	JUL	AUG	SEPT	OCT	NOV	DEC	TOTAL
GRAIN													
GRAIN													
PASTURE													

MILK PRODUCTION

GOAT'S NAME:		IDENTIFICATION:	
BREED:	DATE OF BIRTH:	KIDDING DATE:	

JANUARY		AVERAGE LBS / DAY X 31 DAYS =		LBS
FEBRUARY		AVERAGE LBS / DAY X 31 DAYS =		LBS
MARCH		AVERAGE LBS / DAY X 31 DAYS =		LBS
APRIL		AVERAGE LBS / DAY X 31 DAYS =		LBS
MAY		AVERAGE LBS / DAY X 31 DAYS =		LBS
JUNE		AVERAGE LBS / DAY X 31 DAYS =		LBS
JULY		AVERAGE LBS / DAY X 31 DAYS =		LBS
AUGUST		AVERAGE LBS / DAY X 31 DAYS =		LBS
SEPTEMBER		AVERAGE LBS / DAY X 31 DAYS =		LBS
OCTOBER		AVERAGE LBS / DAY X 31 DAYS =		LBS
NOVEMBER		AVERAGE LBS / DAY X 31 DAYS =		LBS
DECEMBER		AVERAGE LBS / DAY X 31 DAYS =		LBS
YEARLY TOTAL MILK PRODUCED =				LBS

TOTAL VALUE OF MILK PRODUCED FOR THE YEAR

	LBS X $		VALUE PER LBS =	

GOAT INFORMATION

PHOTO

NAME	☐ BUCK	☐ DOE
BREED	BIRTH DATE:	
DATE ACQUIRED:	HOW ACQUIRED: ☐ BORN ON FARM ☐ PURCHASED ☐ LEASED	
COLORS / IDENTIFYING MARKS:		
PURPOSE: ☐ MILK ☐ MEAT ☐ PET ☐ OTHER		

PEDIGREE CHART

- SIRE
  - GRAND SIRE
    - GREAT GRAND SIRE
    - GREAT GRAND DAM
  - GRAND DAM
    - GREAT GRAND SIRE
    - GREAT GRAND DAM
- DAM
  - GRAND SIRE
    - GREAT GRAND SIRE
    - GREAT GRAND DAM
  - GRAND DAM
    - GREAT GRAND SIRE
    - GREAT GRAND DAM

MEDICAL INFORMATION

INJURY OR ILLNESS

DATE	DESCRIPTION OR NATURE OF ILLNESS	TREATMENT

PARASITE CONTROL

DATE	METHOD OR DEWORMER	DATE	METHOD OR DEWORMER

TESTING RECORD

DATE	TEST PERFORMED (CAE, CL, TB…)	RESULT	DATE	TEST PERFORMED (CAE, CL, TB…)	RESULT

INJURY OR ILLNESS

DATE	TARGET DISEASE	DRUG OR SUPPLEMENT USED	DOSAGE	RESULTS

DOE'S KIDDING RECORD

DOE'S NAME:

DATE BREED	KIDDING DATE	# OF KIDS	SEX D/B	NAME OF KID	SIRE OF KID	WEIGHT	TATTOO

BUCK'S RECORD OF PROGENY

DOE'S NAME:	

YEAR	BRED TO	KIDS	DOE/BUCK

GOAT RECORD

GOAT'S NAME:

IDENTIFICATION:

BREED:

DATE OF BIRTH:

DATE OF WEANED:

WEIGHT (POUNDS)													
BIRTH	JAN	FEB	MAR	APR	MAY	JUN	JUL	AUG	SEPT	OCT	NOV	DEC	FINAL

FEED RECORD													
	JAN	FEB	MAR	APR	MAY	JUN	JUL	AUG	SEPT	OCT	NOV	DEC	TOTAL
GRAIN													
GRAIN														
PASTURE													

MILK PRODUCTION

GOAT'S NAME:		IDENTIFICATION:	
BREED:	DATE OF BIRTH:	KIDDING DATE:	

JANUARY		AVERAGE LBS / DAY X 31 DAYS =		LBS
FEBRUARY		AVERAGE LBS / DAY X 31 DAYS =		LBS
MARCH		AVERAGE LBS / DAY X 31 DAYS =		LBS
APRIL		AVERAGE LBS / DAY X 31 DAYS =		LBS
MAY		AVERAGE LBS / DAY X 31 DAYS =		LBS
JUNE		AVERAGE LBS / DAY X 31 DAYS =		LBS
JULY		AVERAGE LBS / DAY X 31 DAYS =		LBS
AUGUST		AVERAGE LBS / DAY X 31 DAYS =		LBS
SEPTEMBER		AVERAGE LBS / DAY X 31 DAYS =		LBS
OCTOBER		AVERAGE LBS / DAY X 31 DAYS =		LBS
NOVEMBER		AVERAGE LBS / DAY X 31 DAYS =		LBS
DECEMBER		AVERAGE LBS / DAY X 31 DAYS =		LBS
YEARLY TOTAL MILK PRODUCED =				LBS
TOTAL VALUE OF MILK PRODUCED FOR THE YEAR				
	LBS X $		VALUE PER LBS =	

GOAT INFORMATION

PHOTO

NAME	☐ BUCK	☐ DOE
BREED	BIRTH DATE:	

DATE ACQUIRED:	HOW ACQUIRED: ☐ BORN ON FARM ☐ PURCHASED ☐ LEASED

COLORS / IDENTIFYING MARKS:

PURPOSE:	☐ MILK	☐ MEAT	☐ PET	☐ OTHER

PEDIGREE CHART

- SIRE
  - GRAND SIRE
    - GREAT GRAND SIRE
    - GREAT GRAND DAM
  - GRAND DAM
    - GREAT GRAND SIRE
    - GREAT GRAND DAM
- DAM
  - GRAND SIRE
    - GREAT GRAND SIRE
    - GREAT GRAND DAM
  - GRAND DAM
    - GREAT GRAND SIRE
    - GREAT GRAND DAM

MEDICAL INFORMATION

INJURY OR ILLNESS

DATE	DESCRIPTION OR NATURE OF ILLNESS	TREATMENT

PARASITE CONTROL

DATE	METHOD OR DEWORMER		DATE	METHOD OR DEWORMER

TESTING RECORD

DATE	TEST PERFORMED (CAE, CL, TB...)	RESULT		DATE	TEST PERFORMED (CAE, CL, TB...)	RESULT

INJURY OR ILLNESS

DATE	TARGET DISEASE	DRUG OR SUPPLEMENT USED	DOSAGE	RESULTS

DOE'S KIDDING RECORD

DOE'S NAME:

DATE BREED	KIDDING DATE	# OF KIDS	SEX D/B	NAME OF KID	SIRE OF KID	WEIGHT	TATTOO

BUCK'S RECORD OF PROGENY

DOE'S NAME:

YEAR	BRED TO	KIDS	DOE/BUCK

GOAT RECORD

GOAT'S NAME:		IDENTIFICATION:	
BREED:	DATE OF BIRTH:		DATE OF WEANED:

WEIGHT (POUNDS)

BIRTH	JAN	FEB	MAR	APR	MAY	JUN	JUL	AUG	SEPT	OCT	NOV	DEC	FINAL

FEED RECORD

	JAN	FEB	MAR	APR	MAY	JUN	JUL	AUG	SEPT	OCT	NOV	DEC	TOTAL
GRAIN													
GRAIN													
PASTURE													

MILK PRODUCTION

GOAT'S NAME:		IDENTIFICATION:	
BREED:	DATE OF BIRTH:	KIDDING DATE:	

JANUARY		AVERAGE LBS / DAY X 31 DAYS =		LBS
FEBRUARY		AVERAGE LBS / DAY X 31 DAYS =		LBS
MARCH		AVERAGE LBS / DAY X 31 DAYS =		LBS
APRIL		AVERAGE LBS / DAY X 31 DAYS =		LBS
MAY		AVERAGE LBS / DAY X 31 DAYS =		LBS
JUNE		AVERAGE LBS / DAY X 31 DAYS =		LBS
JULY		AVERAGE LBS / DAY X 31 DAYS =		LBS
AUGUST		AVERAGE LBS / DAY X 31 DAYS =		LBS
SEPTEMBER		AVERAGE LBS / DAY X 31 DAYS =		LBS
OCTOBER		AVERAGE LBS / DAY X 31 DAYS =		LBS
NOVEMBER		AVERAGE LBS / DAY X 31 DAYS =		LBS
DECEMBER		AVERAGE LBS / DAY X 31 DAYS =		LBS
YEARLY TOTAL MILK PRODUCED =				LBS
TOTAL VALUE OF MILK PRODUCED FOR THE YEAR				
	LBS X $		VALUE PER LBS =	

GOAT INFORMATION

PHOTO

NAME	☐ BUCK	☐ DOE
BREED	BIRTH DATE:	

DATE ACQUIRED:	HOW ACQUIRED: ☐ BORN ON FARM ☐ PURCHASED ☐ LEASED

COLORS / IDENTIFYING MARKS:

PURPOSE:	☐ MILK	☐ MEAT	☐ PET	☐ OTHER

PEDIGREE CHART

SIRE

GRAND SIRE
- GREAT GRAND SIRE
- GREAT GRAND DAM

GRAND DAM
- GREAT GRAND SIRE
- GREAT GRAND DAM

DAM

GRAND SIRE
- GREAT GRAND SIRE
- GREAT GRAND DAM

GRAND DAM
- GREAT GRAND SIRE
- GREAT GRAND DAM

MEDICAL INFORMATION

INJURY OR ILLNESS

DATE	DESCRIPTION OR NATURE OF ILLNESS	TREATMENT

PARASITE CONTROL

DATE	METHOD OR DEWORMER	DATE	METHOD OR DEWORMER

TESTING RECORD

DATE	TEST PERFORMED (CAE, CL, TB...)	RESULT	DATE	TEST PERFORMED (CAE, CL, TB...)	RESULT

INJURY OR ILLNESS

DATE	TARGET DISEASE	DRUG OR SUPPLEMENT USED	DOSAGE	RESULTS

DOE'S KIDDING RECORD

DOE'S NAME:

DATE BREED	KIDDING DATE	# OF KIDS	SEX D/B	NAME OF KID	SIRE OF KID	WEIGHT	TATTOO

BUCK'S RECORD OF PROGENY

DOE'S NAME:	

YEAR	BRED TO	KIDS	DOE/BUCK

GOAT RECORD

GOAT'S NAME:		IDENTIFICATION:	
BREED:	DATE OF BIRTH:		DATE OF WEANED:

WEIGHT (POUNDS)													
BIRTH	**JAN**	**FEB**	**MAR**	**APR**	**MAY**	**JUN**	**JUL**	**AUG**	**SEPT**	**OCT**	**NOV**	**DEC**	**FINAL**

FEED RECORD												
	JAN	**FEB**	**MAR**	**APR**	**MAY**	**JUN**	**JUL**	**AUG**	**SEPT**	**OCT**	**NOV**	**DEC**	**TOTAL**
GRAIN													
GRAIN													
PASTURE													

MILK PRODUCTION

GOAT'S NAME:		IDENTIFICATION:	
BREED:	DATE OF BIRTH:		KIDDING DATE:

JANUARY		AVERAGE LBS / DAY X 31 DAYS =		LBS
FEBRUARY		AVERAGE LBS / DAY X 31 DAYS =		LBS
MARCH		AVERAGE LBS / DAY X 31 DAYS =		LBS
APRIL		AVERAGE LBS / DAY X 31 DAYS =		LBS
MAY		AVERAGE LBS / DAY X 31 DAYS =		LBS
JUNE		AVERAGE LBS / DAY X 31 DAYS =		LBS
JULY		AVERAGE LBS / DAY X 31 DAYS =		LBS
AUGUST		AVERAGE LBS / DAY X 31 DAYS =		LBS
SEPTEMBER		AVERAGE LBS / DAY X 31 DAYS =		LBS
OCTOBER		AVERAGE LBS / DAY X 31 DAYS =		LBS
NOVEMBER		AVERAGE LBS / DAY X 31 DAYS =		LBS
DECEMBER		AVERAGE LBS / DAY X 31 DAYS =		LBS
YEARLY TOTAL MILK PRODUCED =				LBS
TOTAL VALUE OF MILK PRODUCED FOR THE YEAR				
	LBS X $		VALUE PER LBS =	

GOAT INFORMATION

PHOTO

NAME	☐ BUCK	☐ DOE
BREED	BIRTH DATE:	
DATE ACQUIRED:	HOW ACQUIRED: ☐ BORN ON FARM ☐ PURCHASED ☐ LEASED	
COLORS / IDENTIFYING MARKS:		
PURPOSE: ☐ MILK ☐ MEAT ☐ PET ☐ OTHER		

PEDIGREE CHART

GREAT GRAND SIRE

GRAND SIRE

GREAT GRAND DAM

SIRE

GREAT GRAND SIRE

GRAND DAM

GREAT GRAND DAM

GREAT GRAND SIRE

GRAND SIRE

GREAT GRAND DAM

DAM

GREAT GRAND SIRE

GRAND DAM

GREAT GRAND DAM

MEDICAL INFORMATION

INJURY OR ILLNESS

DATE	DESCRIPTION OR NATURE OF ILLNESS	TREATMENT

PARASITE CONTROL

DATE	METHOD OR DEWORMER	DATE	METHOD OR DEWORMER

TESTING RECORD

DATE	TEST PERFORMED (CAE, CL, TB...)	RESULT	DATE	TEST PERFORMED (CAE, CL, TB...)	RESULT

INJURY OR ILLNESS

DATE	TARGET DISEASE	DRUG OR SUPPLEMENT USED	DOSAGE	RESULTS

DOE'S KIDDING RECORD

DOE'S NAME:

DATE BREED	KIDDING DATE	# OF KIDS	SEX D/B	NAME OF KID	SIRE OF KID	WEIGHT	TATTOO

BUCK'S RECORD OF PROGENY

DOE'S NAME:

YEAR	BRED TO	KIDS	DOE/BUCK

GOAT RECORD

GOAT'S NAME:		IDENTIFICATION:	
BREED:	DATE OF BIRTH:	DATE OF WEANED:	

WEIGHT (POUNDS)

BIRTH	JAN	FEB	MAR	APR	MAY	JUN	JUL	AUG	SEPT	OCT	NOV	DEC	FINAL

FEED RECORD

	JAN	FEB	MAR	APR	MAY	JUN	JUL	AUG	SEPT	OCT	NOV	DEC	TOTAL
GRAIN													
GRAIN													
PASTURE													

MILK PRODUCTION

GOAT'S NAME:		IDENTIFICATION:		
BREED:		DATE OF BIRTH:	KIDDING DATE:	

JANUARY		AVERAGE LBS / DAY X 31 DAYS =		LBS
FEBRUARY		AVERAGE LBS / DAY X 31 DAYS =		LBS
MARCH		AVERAGE LBS / DAY X 31 DAYS =		LBS
APRIL		AVERAGE LBS / DAY X 31 DAYS =		LBS
MAY		AVERAGE LBS / DAY X 31 DAYS =		LBS
JUNE		AVERAGE LBS / DAY X 31 DAYS =		LBS
JULY		AVERAGE LBS / DAY X 31 DAYS =		LBS
AUGUST		AVERAGE LBS / DAY X 31 DAYS =		LBS
SEPTEMBER		AVERAGE LBS / DAY X 31 DAYS =		LBS
OCTOBER		AVERAGE LBS / DAY X 31 DAYS =		LBS
NOVEMBER		AVERAGE LBS / DAY X 31 DAYS =		LBS
DECEMBER		AVERAGE LBS / DAY X 31 DAYS =		LBS
YEARLY TOTAL MILK PRODUCED =				LBS
TOTAL VALUE OF MILK PRODUCED FOR THE YEAR				
	LBS X $		VALUE PER LBS =	

GOAT INFORMATION

PHOTO

NAME		☐ BUCK	☐ DOE
BREED		BIRTH DATE:	
DATE ACQUIRED:	HOW ACQUIRED: ☐ BORN ON FARM ☐ PURCHASED ☐ LEASED		
COLORS / IDENTIFYING MARKS:			
PURPOSE: ☐ MILK ☐ MEAT ☐ PET ☐ OTHER			

PEDIGREE CHART

			GREAT GRAND SIRE
		GRAND SIRE	
	SIRE		GREAT GRAND DAM
			GREAT GRAND SIRE
		GRAND DAM	
			GREAT GRAND DAM
			GREAT GRAND SIRE
		GRAND SIRE	
	DAM		GREAT GRAND DAM
			GREAT GRAND SIRE
		GRAND DAM	
			GREAT GRAND DAM

MEDICAL INFORMATION

INJURY OR ILLNESS

DATE	DESCRIPTION OR NATURE OF ILLNESS	TREATMENT

PARASITE CONTROL

DATE	METHOD OR DEWORMER	DATE	METHOD OR DEWORMER

TESTING RECORD

DATE	TEST PERFORMED (CAE, CL, TB...)	RESULT	DATE	TEST PERFORMED (CAE, CL, TB...)	RESULT

INJURY OR ILLNESS

DATE	TARGET DISEASE	DRUG OR SUPPLEMENT USED	DOSAGE	RESULTS

DOE'S KIDDING RECORD

DOE'S NAME:

DATE BREED	KIDDING DATE	# OF KIDS	SEX D/B	NAME OF KID	SIRE OF KID	WEIGHT	TATTOO

BUCK'S RECORD OF PROGENY

DOE'S NAME:	

YEAR	BRED TO	KIDS	DOE/BUCK

GOAT RECORD

GOAT'S NAME:		IDENTIFICATION:	
BREED:	DATE OF BIRTH:	DATE OF WEANED:	

WEIGHT (POUNDS)

BIRTH	JAN	FEB	MAR	APR	MAY	JUN	JUL	AUG	SEPT	OCT	NOV	DEC	FINAL

FEED RECORD

	JAN	FEB	MAR	APR	MAY	JUN	JUL	AUG	SEPT	OCT	NOV	DEC	TOTAL
GRAIN													
GRAIN													
PASTURE													

MILK PRODUCTION

GOAT'S NAME:		IDENTIFICATION:	
BREED:	DATE OF BIRTH:	KIDDING DATE:	

JANUARY		AVERAGE LBS / DAY X 31 DAYS =		LBS
FEBRUARY		AVERAGE LBS / DAY X 31 DAYS =		LBS
MARCH		AVERAGE LBS / DAY X 31 DAYS =		LBS
APRIL		AVERAGE LBS / DAY X 31 DAYS =		LBS
MAY		AVERAGE LBS / DAY X 31 DAYS =		LBS
JUNE		AVERAGE LBS / DAY X 31 DAYS =		LBS
JULY		AVERAGE LBS / DAY X 31 DAYS =		LBS
AUGUST		AVERAGE LBS / DAY X 31 DAYS =		LBS
SEPTEMBER		AVERAGE LBS / DAY X 31 DAYS =		LBS
OCTOBER		AVERAGE LBS / DAY X 31 DAYS =		LBS
NOVEMBER		AVERAGE LBS / DAY X 31 DAYS =		LBS
DECEMBER		AVERAGE LBS / DAY X 31 DAYS =		LBS
YEARLY TOTAL MILK PRODUCED =				LBS

TOTAL VALUE OF MILK PRODUCED FOR THE YEAR

	LBS X $		VALUE PER LBS =	

GOAT INFORMATION

PHOTO

NAME		☐ BUCK	☐ DOE
BREED		BIRTH DATE:	
DATE ACQUIRED:	HOW ACQUIRED: ☐ BORN ON FARM ☐ PURCHASED ☐ LEASED		
COLORS / IDENTIFYING MARKS:			
PURPOSE: ☐ MILK ☐ MEAT ☐ PET ☐ OTHER			

PEDIGREE CHART

SIRE

GRAND SIRE

GREAT GRAND SIRE

GREAT GRAND DAM

GRAND DAM

GREAT GRAND SIRE

GREAT GRAND DAM

DAM

GRAND SIRE

GREAT GRAND SIRE

GREAT GRAND DAM

GRAND DAM

GREAT GRAND SIRE

GREAT GRAND DAM

MEDICAL INFORMATION

INJURY OR ILLNESS

DATE	DESCRIPTION OR NATURE OF ILLNESS	TREATMENT

PARASITE CONTROL

DATE	METHOD OR DEWORMER	DATE	METHOD OR DEWORMER

TESTING RECORD

DATE	TEST PERFORMED (CAE, CL, TB...)	RESULT	DATE	TEST PERFORMED (CAE, CL, TB...)	RESULT

INJURY OR ILLNESS

DATE	TARGET DISEASE	DRUG OR SUPPLEMENT USED	DOSAGE	RESULTS

DOE'S KIDDING RECORD

DOE'S NAME:

DATE BREED	KIDDING DATE	# OF KIDS	SEX D/B	NAME OF KID	SIRE OF KID	WEIGHT	TATTOO

BUCK'S RECORD OF PROGENY

DOE'S NAME:	

YEAR	BRED TO	KIDS	DOE/BUCK

GOAT RECORD

GOAT'S NAME:		IDENTIFICATION:	
BREED:	DATE OF BIRTH:		DATE OF WEANED:

WEIGHT (POUNDS)													
BIRTH	**JAN**	**FEB**	**MAR**	**APR**	**MAY**	**JUN**	**JUL**	**AUG**	**SEPT**	**OCT**	**NOV**	**DEC**	**FINAL**

FEED RECORD													
	JAN	**FEB**	**MAR**	**APR**	**MAY**	**JUN**	**JUL**	**AUG**	**SEPT**	**OCT**	**NOV**	**DEC**	**TOTAL**
GRAIN													
GRAIN													
PASTURE													

MILK PRODUCTION

GOAT'S NAME:		IDENTIFICATION:	
BREED:	DATE OF BIRTH:	KIDDING DATE:	

JANUARY		AVERAGE LBS / DAY X 31 DAYS =		LBS
FEBRUARY		AVERAGE LBS / DAY X 31 DAYS =		LBS
MARCH		AVERAGE LBS / DAY X 31 DAYS =		LBS
APRIL		AVERAGE LBS / DAY X 31 DAYS =		LBS
MAY		AVERAGE LBS / DAY X 31 DAYS =		LBS
JUNE		AVERAGE LBS / DAY X 31 DAYS =		LBS
JULY		AVERAGE LBS / DAY X 31 DAYS =		LBS
AUGUST		AVERAGE LBS / DAY X 31 DAYS =		LBS
SEPTEMBER		AVERAGE LBS / DAY X 31 DAYS =		LBS
OCTOBER		AVERAGE LBS / DAY X 31 DAYS =		LBS
NOVEMBER		AVERAGE LBS / DAY X 31 DAYS =		LBS
DECEMBER		AVERAGE LBS / DAY X 31 DAYS =		LBS
YEARLY TOTAL MILK PRODUCED =				LBS
TOTAL VALUE OF MILK PRODUCED FOR THE YEAR				
	LBS X $		VALUE PER LBS =	

GOAT INFORMATION

PHOTO

NAME	☐ BUCK	☐ DOE
BREED	BIRTH DATE:	
DATE ACQUIRED:	HOW ACQUIRED: ☐ BORN ON FARM ☐ PURCHASED ☐ LEASED	
COLORS / IDENTIFYING MARKS:		
PURPOSE: ☐ MILK ☐ MEAT ☐ PET ☐ OTHER		

PEDIGREE CHART

			GREAT GRAND SIRE
		GRAND SIRE	
			GREAT GRAND DAM
	SIRE		
			GREAT GRAND SIRE
		GRAND DAM	
			GREAT GRAND DAM
			GREAT GRAND SIRE
		GRAND SIRE	
			GREAT GRAND DAM
	DAM		
			GREAT GRAND SIRE
		GRAND DAM	
			GREAT GRAND DAM

MEDICAL INFORMATION

INJURY OR ILLNESS

DATE	DESCRIPTION OR NATURE OF ILLNESS	TREATMENT

PARASITE CONTROL

DATE	METHOD OR DEWORMER	DATE	METHOD OR DEWORMER

TESTING RECORD

DATE	TEST PERFORMED (CAE, CL, TB…)	RESULT	DATE	TEST PERFORMED (CAE, CL, TB…)	RESULT

INJURY OR ILLNESS

DATE	TARGET DISEASE	DRUG OR SUPPLEMENT USED	DOSAGE	RESULTS

DOE'S KIDDING RECORD

DOE'S NAME:

DATE BREED	KIDDING DATE	# OF KIDS	SEX D/B	NAME OF KID	SIRE OF KID	WEIGHT	TATTOO

BUCK'S RECORD OF PROGENY

DOE'S NAME:			

YEAR	BRED TO	KIDS	DOE/BUCK

GOAT RECORD

GOAT'S NAME:		IDENTIFICATION:	
BREED:	DATE OF BIRTH:		DATE OF WEANED:

WEIGHT (POUNDS)

BIRTH	JAN	FEB	MAR	APR	MAY	JUN	JUL	AUG	SEPT	OCT	NOV	DEC	FINAL

FEED RECORD

	JAN	FEB	MAR	APR	MAY	JUN	JUL	AUG	SEPT	OCT	NOV	DEC	TOTAL
GRAIN													
GRAIN													
PASTURE													

MILK PRODUCTION

GOAT'S NAME:		IDENTIFICATION:		
BREED:		DATE OF BIRTH:		KIDDING DATE:

JANUARY		AVERAGE LBS / DAY X 31 DAYS =		LBS
FEBRUARY		AVERAGE LBS / DAY X 31 DAYS =		LBS
MARCH		AVERAGE LBS / DAY X 31 DAYS =		LBS
APRIL		AVERAGE LBS / DAY X 31 DAYS =		LBS
MAY		AVERAGE LBS / DAY X 31 DAYS =		LBS
JUNE		AVERAGE LBS / DAY X 31 DAYS =		LBS
JULY		AVERAGE LBS / DAY X 31 DAYS =		LBS
AUGUST		AVERAGE LBS / DAY X 31 DAYS =		LBS
SEPTEMBER		AVERAGE LBS / DAY X 31 DAYS =		LBS
OCTOBER		AVERAGE LBS / DAY X 31 DAYS =		LBS
NOVEMBER		AVERAGE LBS / DAY X 31 DAYS =		LBS
DECEMBER		AVERAGE LBS / DAY X 31 DAYS =		LBS
YEARLY TOTAL MILK PRODUCED =				LBS
TOTAL VALUE OF MILK PRODUCED FOR THE YEAR				
	LBS X $		VALUE PER LBS =	

GOAT INFORMATION

PHOTO

NAME		☐ BUCK	☐ DOE
BREED		BIRTH DATE:	
DATE ACQUIRED:	HOW ACQUIRED: ☐ BORN ON FARM ☐ PURCHASED ☐ LEASED		
COLORS / IDENTIFYING MARKS:			
PURPOSE: ☐ MILK ☐ MEAT ☐ PET ☐ OTHER			

PEDIGREE CHART

SIRE

GRAND SIRE

GREAT GRAND SIRE

GREAT GRAND DAM

GRAND DAM

GREAT GRAND SIRE

GREAT GRAND DAM

DAM

GRAND SIRE

GREAT GRAND SIRE

GREAT GRAND DAM

GRAND DAM

GREAT GRAND SIRE

GREAT GRAND DAM

MEDICAL INFORMATION

INJURY OR ILLNESS

DATE	DESCRIPTION OR NATURE OF ILLNESS	TREATMENT

PARASITE CONTROL

DATE	METHOD OR DEWORMER	DATE	METHOD OR DEWORMER

TESTING RECORD

DATE	TEST PERFORMED (CAE, CL, TB…)	RESULT	DATE	TEST PERFORMED (CAE, CL, TB…)	RESULT

INJURY OR ILLNESS

DATE	TARGET DISEASE	DRUG OR SUPPLEMENT USED	DOSAGE	RESULTS

DOE'S KIDDING RECORD

DOE'S NAME:	

DATE BREED	KIDDING DATE	# OF KIDS	SEX D/B	NAME OF KID	SIRE OF KID	WEIGHT	TATTOO

BUCK'S RECORD OF PROGENY

DOE'S NAME:

YEAR	BRED TO	KIDS	DOE/BUCK

GOAT RECORD

GOAT'S NAME:

IDENTIFICATION:

BREED:

DATE OF BIRTH:

DATE OF WEANED:

WEIGHT (POUNDS)

BIRTH	JAN	FEB	MAR	APR	MAY	JUN	JUL	AUG	SEPT	OCT	NOV	DEC	FINAL

FEED RECORD

	JAN	FEB	MAR	APR	MAY	JUN	JUL	AUG	SEPT	OCT	NOV	DEC	TOTAL
GRAIN													
GRAIN													
PASTURE													

MILK PRODUCTION

GOAT'S NAME:

IDENTIFICATION:

BREED:

DATE OF BIRTH:

KIDDING DATE:

JANUARY		AVERAGE LBS / DAY X 31 DAYS =		LBS
FEBRUARY		AVERAGE LBS / DAY X 31 DAYS =		LBS
MARCH		AVERAGE LBS / DAY X 31 DAYS =		LBS
APRIL		AVERAGE LBS / DAY X 31 DAYS =		LBS
MAY		AVERAGE LBS / DAY X 31 DAYS =		LBS
JUNE		AVERAGE LBS / DAY X 31 DAYS =		LBS
JULY		AVERAGE LBS / DAY X 31 DAYS =		LBS
AUGUST		AVERAGE LBS / DAY X 31 DAYS =		LBS
SEPTEMBER		AVERAGE LBS / DAY X 31 DAYS =		LBS
OCTOBER		AVERAGE LBS / DAY X 31 DAYS =		LBS
NOVEMBER		AVERAGE LBS / DAY X 31 DAYS =		LBS
DECEMBER		AVERAGE LBS / DAY X 31 DAYS =		LBS
YEARLY TOTAL MILK PRODUCED =				LBS

TOTAL VALUE OF MILK PRODUCED FOR THE YEAR

	LBS X $		VALUE PER LBS =	

GOAT INFORMATION

PHOTO

NAME	☐ BUCK	☐ DOE
BREED	BIRTH DATE:	
DATE ACQUIRED:	HOW ACQUIRED: ☐ BORN ON FARM ☐ PURCHASED ☐ LEASED	
COLORS / IDENTIFYING MARKS:		
PURPOSE: ☐ MILK ☐ MEAT ☐ PET ☐ OTHER		

PEDIGREE CHART

			GREAT GRAND SIRE
		GRAND SIRE	
			GREAT GRAND DAM
	SIRE		
			GREAT GRAND SIRE
		GRAND DAM	
			GREAT GRAND DAM
			GREAT GRAND SIRE
		GRAND SIRE	
			GREAT GRAND DAM
	DAM		
			GREAT GRAND SIRE
		GRAND DAM	
			GREAT GRAND DAM

MEDICAL INFORMATION

INJURY OR ILLNESS

DATE	DESCRIPTION OR NATURE OF ILLNESS	TREATMENT

PARASITE CONTROL

DATE	METHOD OR DEWORMER	DATE	METHOD OR DEWORMER

TESTING RECORD

DATE	TEST PERFORMED (CAE, CL, TB...)	RESULT	DATE	TEST PERFORMED (CAE, CL, TB...)	RESULT

INJURY OR ILLNESS

DATE	TARGET DISEASE	DRUG OR SUPPLEMENT USED	DOSAGE	RESULTS

DOE'S KIDDING RECORD

DOE'S NAME:	

DATE BREED	KIDDING DATE	# OF KIDS	SEX D/B	NAME OF KID	SIRE OF KID	WEIGHT	TATTOO

BUCK'S RECORD OF PROGENY

DOE'S NAME:

YEAR	BRED TO	KIDS	DOE/BUCK

GOAT RECORD

GOAT'S NAME:		IDENTIFICATION:	
BREED:	DATE OF BIRTH:		DATE OF WEANED:

WEIGHT (POUNDS)

BIRTH	JAN	FEB	MAR	APR	MAY	JUN	JUL	AUG	SEPT	OCT	NOV	DEC	FINAL

FEED RECORD

	JAN	FEB	MAR	APR	MAY	JUN	JUL	AUG	SEPT	OCT	NOV	DEC	TOTAL
GRAIN													
GRAIN													
PASTURE													

MILK PRODUCTION

GOAT'S NAME:		IDENTIFICATION:	
BREED:	DATE OF BIRTH:	KIDDING DATE:	

JANUARY		AVERAGE LBS / DAY X 31 DAYS =		LBS
FEBRUARY		AVERAGE LBS / DAY X 31 DAYS =		LBS
MARCH		AVERAGE LBS / DAY X 31 DAYS =		LBS
APRIL		AVERAGE LBS / DAY X 31 DAYS =		LBS
MAY		AVERAGE LBS / DAY X 31 DAYS =		LBS
JUNE		AVERAGE LBS / DAY X 31 DAYS =		LBS
JULY		AVERAGE LBS / DAY X 31 DAYS =		LBS
AUGUST		AVERAGE LBS / DAY X 31 DAYS =		LBS
SEPTEMBER		AVERAGE LBS / DAY X 31 DAYS =		LBS
OCTOBER		AVERAGE LBS / DAY X 31 DAYS =		LBS
NOVEMBER		AVERAGE LBS / DAY X 31 DAYS =		LBS
DECEMBER		AVERAGE LBS / DAY X 31 DAYS =		LBS
YEARLY TOTAL MILK PRODUCED =				LBS
TOTAL VALUE OF MILK PRODUCED FOR THE YEAR				
	LBS X $		VALUE PER LBS =	

GOAT INFORMATION

PHOTO

NAME		☐ BUCK	☐ DOE
BREED		BIRTH DATE:	
DATE ACQUIRED:	HOW ACQUIRED: ☐ BORN ON FARM ☐ PURCHASED ☐ LEASED		
COLORS / IDENTIFYING MARKS:			
PURPOSE: ☐ MILK ☐ MEAT ☐ PET ☐ OTHER			

PEDIGREE CHART

SIRE

GRAND SIRE

GREAT GRAND SIRE

GREAT GRAND DAM

GRAND DAM

GREAT GRAND SIRE

GREAT GRAND DAM

DAM

GRAND SIRE

GREAT GRAND SIRE

GREAT GRAND DAM

GRAND DAM

GREAT GRAND SIRE

GREAT GRAND DAM

MEDICAL INFORMATION

INJURY OR ILLNESS

DATE	DESCRIPTION OR NATURE OF ILLNESS	TREATMENT

PARASITE CONTROL

DATE	METHOD OR DEWORMER	DATE	METHOD OR DEWORMER

TESTING RECORD

DATE	TEST PERFORMED (CAE, CL, TB...)	RESULT	DATE	TEST PERFORMED (CAE, CL, TB...)	RESULT

INJURY OR ILLNESS

DATE	TARGET DISEASE	DRUG OR SUPPLEMENT USED	DOSAGE	RESULTS

DOE'S KIDDING RECORD

DOE'S NAME:	

DATE BREED	KIDDING DATE	# OF KIDS	SEX D/B	NAME OF KID	SIRE OF KID	WEIGHT	TATTOO

BUCK'S RECORD OF PROGENY

DOE'S NAME:	

YEAR	BRED TO	KIDS	DOE/BUCK

GOAT RECORD

GOAT'S NAME:		IDENTIFICATION:	
BREED:	DATE OF BIRTH:		DATE OF WEANED:

WEIGHT (POUNDS)

BIRTH	JAN	FEB	MAR	APR	MAY	JUN	JUL	AUG	SEPT	OCT	NOV	DEC	FINAL

FEED RECORD

	JAN	FEB	MAR	APR	MAY	JUN	JUL	AUG	SEPT	OCT	NOV	DEC	TOTAL
GRAIN													
GRAIN													
PASTURE													

MILK PRODUCTION

GOAT'S NAME:		IDENTIFICATION:	
BREED:	**DATE OF BIRTH:**	**KIDDING DATE:**	

JANUARY		AVERAGE LBS / DAY X 31 DAYS =		LBS
FEBRUARY		AVERAGE LBS / DAY X 31 DAYS =		LBS
MARCH		AVERAGE LBS / DAY X 31 DAYS =		LBS
APRIL		AVERAGE LBS / DAY X 31 DAYS =		LBS
MAY		AVERAGE LBS / DAY X 31 DAYS =		LBS
JUNE		AVERAGE LBS / DAY X 31 DAYS =		LBS
JULY		AVERAGE LBS / DAY X 31 DAYS =		LBS
AUGUST		AVERAGE LBS / DAY X 31 DAYS =		LBS
SEPTEMBER		AVERAGE LBS / DAY X 31 DAYS =		LBS
OCTOBER		AVERAGE LBS / DAY X 31 DAYS =		LBS
NOVEMBER		AVERAGE LBS / DAY X 31 DAYS =		LBS
DECEMBER		AVERAGE LBS / DAY X 31 DAYS =		LBS
YEARLY TOTAL MILK PRODUCED =				LBS

TOTAL VALUE OF MILK PRODUCED FOR THE YEAR

	LBS X $		VALUE PER LBS =	

NOTES